This book provides a thorough examination of the use of elasticity in solving geotechnical engineering problems in a style that will be accessible to upper level students in civil engineering, geological engineering, and of the earth sciences. The first two chapters present a basic framework of the theory of elasticity and describe test procedures for determination of elastic parameters for soils. Chapters 3 and 4 present the fundamental solutions of Boussinesq, Kelvin, Cerrutti, and Mindlin, and use these to formulate solutions to problems of practical interest in geotechnical engineering. The book concludes with a sequence of appendixes designed to provide the interested student with details of the theory of elasticity that would be of considerable assistance to a deeper understanding of the main text.

ELASTICITY AND GEOMECHANICS

ELASTICITY AND GEOMECHANICS

R. O. DAVIS
University of Canterbury

A. P. S. SELVADURAI
McGill University

CAMBRIDGE
UNIVERSITY PRESS

CAMBRIDGE UNIVERSITY PRESS
Cambridge, New York, Melbourne, Madrid, Cape Town, Singapore,
São Paulo, Delhi, Dubai, Tokyo

Cambridge University Press
The Edinburgh Building, Cambridge CB2 8RU, UK

Published in the United States of America by Cambridge University Press, New York

www.cambridge.org
Information on this title: www.cambridge.org/9780521498272

First published 1996

A catalogue record for this publication is available from the British Library

Library of Congress Cataloguing in Publication data
Davis, R. O.
Elasticity and geomechanics / R. O. Davis, A. P. S. Selvadurai.
p. cm.
Includes bibliographical references.
ISBN 0-521-49506-7 (hc). – ISBN 0-521-49827-9 (pbk.)
1. Engineering geology – Mathematics. 2. Elasticity.
I. Selvadurai, A. P. S. II. Title.
TA705.D3 1996
624.1′51 – dc20   95-17507
                                   CIP

ISBN 978-0-521-49506-6 Hardback
ISBN 978-0-521-49827-2 Paperback

Transferred to digital printing 2009

Contents

Preface

The linear theory of elasticity has enjoyed a fairly long and profitable history in the field of foundation engineering. Geotechnical engineers have turned to elasticity for answers to a variety of questions, and despite the sure knowledge that the answers are at best approximate, they will continue to do so for some time. The reasons for this lie in the essential simplicity of the relevant elastic solutions. The point load solutions of Boussinesq and others, as well as the distributed load solutions that derive from them, can provide answers with only a few lines of calculations. This can be a significant advantage when compared with the time and effort involved in obtaining numerical solutions that employ one of the multitude of existing plasticity models for soil. Furthermore, quite a lot of information concerning soil properties is required to run any of the various plasticity models. In many practical situations, the information may simply be unavailable, and the geotechnical engineer is left with few alternatives. In contrast, elasticity solutions will generally require only a value for the soil modulus and Poisson's ratio, and if the values given are known to be rough approximations, then at least the solution method is in keeping with the input data.

This book is about the use of elasticity in solving geotechnical engineering problems. It is directed toward upper level students in civil engineering or engineering geology. It was motivated when, in 1992, Patrick Selvadurai visited the University of Canterbury. There he found a course taught by Rob Davis, similar to a course Patrick himself taught at Carleton University. Both courses had evolved independently but were surprisingly similar in content, covering some basic applications of elasticity theory in geotechnical engineering.

The book grew from lecture notes and is intended solely as a teaching tool. It is not exhaustive in content but is sufficiently complete to provide a grounding in the linear theory of elasticity, together with an understanding of applications in foundation engineering. The presentation is deliberately informal and conversational in tone. From the standpoint of a student, it should not be

ix

intimidating, but should hopefully put a variety of new ideas into an accessible format.

Four chapters form the main body of the book. Chapter 1 presents a framework of basic ideas from the linear theory of elasticity; deformation, strain and stress, equilibrium and compatibility, and the formulation of problems. Chapter 2 delves into Hooke's law and the elastic constants. The first half of this chapter describes the elastic constants and explains their relation to the material in Chapter 1, while the second half discusses how elastic constants may be determined in a geotechnical context. In Chapter 3 the point load problems of Boussinesq, Kelvin, Cerrutti, and Mindlin, as well as the line load problem of Flamant, are described. The solutions to these problems are presented but techniques for finding solutions are not discussed. The thrust of the book is not to provide another elasticity text. The point load problems are fundamental to geotechnical applications and while their solutions are examined, the corresponding solution methods are of only marginal interest. Chapter 4 uses the solutions from Chapter 3 to consider some basic problems in foundation engineering. The development progresses from consideration of a simple uniformly loaded region on the surface of a homogeneous elastic half-space to more challenging problems involving nonuniform loads, rigid foundations, and layered half-spaces. Some remarks concerning consolidation and in situ testing complete this chapter. Finally, a sequence of appendixes bring the book to a close. These are an important component of the book, designed to provide the interested student with details of elasticity theory that are peripheral to the main text. The compatibility conditions, development of the stress tensor, Saint-Venant's principle, uniqueness, the virtual work principle, and reciprocity relations are all considered in the appendixes.

The book draws together material from a range of sources. The description of the linear theory of elasticity is phrased in a physical rather than purely mathematical context. While mathematical formulations are not avoided, they are construed as the end result of physical thought processes. Examples are used to illustrate important concepts, and a set of problems is placed at the end of each of the four chapters. The book does not directly compete with any other text. While nearly all undergraduate soil mechanics texts contain some elastic analyses, none are devoted to elastic theory. Two other books, Poulos and Davis' *Elastic Solutions in Soil and Rock Mechanics* and Selvadurai's *Elastic Analysis of Soil-Foundation Interaction*, both involve geotechnical uses of elasticity; but the former is a catalog of solutions, while the latter is a treatise. Neither would easily serve as a text for undergraduate students.

Finally, we would like to express our sincere appreciation to two people whose assistance made preparation of the book a far more pleasant task than it might have been. First, all the text and equations were typed with remarkable care and accuracy by Mrs. Pat Roberts. Second, the figures were skillfully drawn by Mrs. Val Grey. To both these people we offer our thanks for a job well done.

R. O. Davis   A. P. S. Selvadurai
Christchurch   Montreal
July 1995   July 1995

1

Some ideas from the theory of elasticity

1.1 Introduction

In this book we will use the linearized theory of elasticity to obtain solutions to some specialized problems in soil mechanics and foundation engineering. For beginning students in geotechnical engineering this may seem like a strange objective. One of the first things we learn about soil response is that, in general, it is neither linear nor elastic. The stress-strain behavior of soil is usually found to be hysteretic and highly nonlinear. It seems quite reasonable then to raise the question of how linear elasticity can be profitably used in relation to soils.

The answer to this question is two-fold. First, elasticity will be a convenient tool, and, in the busy and pressurized world of geotechnical engineering, convenient, reliable, and speedy results are a very positive advantage. The linear theory of elasticity possesses a long history, which has been distinguished not only by firm mathematical foundations but also by the solution of a large number of useful practical problems. Some of these solutions are particularly well-suited to the types of loadings and geometries we encounter in foundation engineering. To have a ready-made solution that either can be used immediately, or with only minor modification, is clearly a significant advantage. However, if the only advantages the theory of elasticity had to offer were convenience and speed, then it would remain an unused tool in the geotechnical engineer's tool kit. The second reason we use elastic theory is because most soils will behave approximately like a linear elastic material, provided the stresses they are subjected to are relatively *small*. By small, we mean the level of shearing stress within the soil deposit is considerably lower than its ultimate strength. This is often the case in many problems in foundation engineering. Usually foundations are designed with factors of safety of three or more. This suggests the stress level, in a general sense, is about one-third of the ultimate soil strength. If we consider a typical stress-strain curve for a fine-grained soil, such as illustrated in Figure 1.1, we might well con-

1

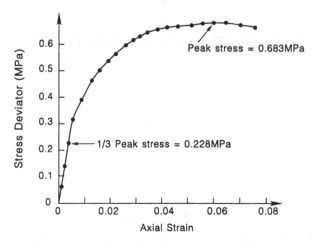

Figure 1.1 Stress deviator versus axial strain taken from results of an actual drained, stress-controlled triaxial test on a uniform beach sand. Note linearity of initial stress-strain response.

clude the behavior is at least approximately linear over the lower one-third of the peak stress range. Thus, for problems in which the soil is not subjected to stresses larger than about one-third the ultimate strength, elastic theory is not only convenient but also offers a rational approximation to the load-deformation behavior we may expect.

There are two other attractions in geotechnical engineering associated with elastic analysis. Elasticity solutions will often give insight into the "mechanics" of a problem which might not be available using other methods. An understanding of elasticity and its use in geotechnical problems frequently has led to new and innovative solutions to old problems. This is particularly true in certain fields such as in-situ testing of soils. The second attraction elasticity offers is that we may use elastic solutions as a check on more sophisticated computer-based solutions. Computational methods can offer an extremely wide range of problem solutions, but it is always comforting for the computer user to see that the program utilized gives correct answers to elastic problems for which the exact solution is known a priori.

The types of geotechnical problems where elasticity will be useful are mostly confined to foundations of structures. Some typical examples are illustrated in Figure 1.2. A great many structures are founded on reinforced concrete footings or pads buried at relatively shallow depths beneath the ground surface. For these types of structures, elastic theory is well-suited to yield estimates for both the stresses in the foundation soil and the displacements or

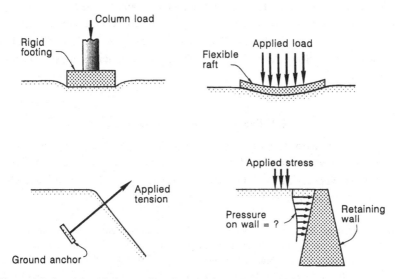

Figure 1.2 Some foundation engineering problems that can be fruitfully tackled using elastic analysis.

settlement of the structure itself. Elasticity is also a powerful tool in assessing the way in which the structure interacts with the soil. Since both the structure and the soil are deformable, but may have different stiffnesses, this problem, referred to as the soil-structure interaction problem, leads to many interesting results.

Practically all text books on foundation engineering contain some material based on elastic analysis. There are two books, however, which are of particular interest. The first is a book written by Harry Poulos and the late Ted Davis called *Elastic Solutions in Soil and Rock Mechanics*, published in 1974. This book is a compendium of solutions gathered from various sources and cataloged for easy reference. The second book is by Patrick Selvadurai, titled *Elastic Analysis of Soil-Foundation Interaction*, published in 1979. This is an advanced text giving a very complete description of the soil-structure interaction problem with formulations and solutions to many problems of practical interest. Both of these books are excellent sources for anyone interested in using linear elasticity in geotechnical engineering. However, both are different from the book you are now holding. This book is designed to introduce both undergraduate and graduate students to the subject. It is not exhaustive in content, and hopefully not exhausting to read. It will touch on a few interesting problems, but omit others. And it will give only a cursory description of the development of elasticity theory. Where we think a slightly more detailed exposition may be of interest to the more mathematically minded reader, it will be presented in Appendixes. In this text we primarily want to concentrate on how elasticity can be used rather than on why it came about. Hence, the title of this chapter, "Some Ideas From the Theory of Elasticity," means exactly what it says.

1.2 The notion of a continuum

The linearized theory of elasticity is a branch of a larger discipline called continuum mechanics. A very lucid definition of continuum mechanics was given by A. J. M. Spencer in his 1966 Inaugural Lecture as Professor of Theoretical Mechanics at the University of Nottingham.

Continuum mechanics means mechanics of continuous media, and a continuous medium is a medium which occupies every point of a continuous region of space. In continuum mechanics we treat our materials – gases, liquids and solids – as though they completely fill up the regions they occupy, with no holes or gaps. This is not what materials are really like. They are made of molecules, and atoms, which in turn are made of smaller particles, and even these particles are not particles in the sense of being little hard lumps of matter. By far the greater part of any piece of any material is empty, and on the atomic scale the chance of any particular point being occupied by matter is small. Continuum mechanics pretends to ignore all this, and in effect smears the atoms, molecules and so on smoothly and uniformly over the region which we suppose our material to occupy. In doing this, of course, we are once again introducing a mathematical idealization; we are replacing a rather complicated idea of what materials are by a less accurate but much simpler one in much the same way as we make an idealization when we treat stars and planets as point masses when making astronomical calculations. There are two justifications for making this idealization. The first and most important is that it works. We can perform certain experiments on a material; on the basis of theory and these experiments we can make predictions about how the material will behave in other experiments....

Let us now examine the relevance of the notion of a continuum in the context of a soil. If we were to divide soil into blocks of one meter on a side we would intuitively expect a typical block to represent the constitution of the soil in some average sense. If we then continued to cut the block into smaller and smaller parts we would quickly find a point at which one part might consist entirely of a particular soil mineral such as quartz, while another part might consist entirely of water, or for that matter, another mineral such as feldspar. So technically a soil can never be a continuum, but this is not a serious obstacle. What we must do in order to treat soil as a continuum, or as Spencer remarked "to make it work," is to agree to apply our results only to volumes sufficiently large enough to encompass significant numbers of soil particles. This will clearly be the case for any reasonable problem in foundation engineering where the characteristic dimension of a foundation will be of the order of meters, whereas the characteristic dimension of a soil could range from 2 μm (e.g., a clay) to 50 mm (gravel).

One consequence of dealing with a continuum is that we can use the concept of a limit. For example, in a continuum the *mass density* ρ is defined as a limit

$$\rho = \lim_{\Delta V \to 0} \frac{\Delta M}{\Delta V}$$

where ΔV is a volume and ΔM the mass of material contained in ΔV. In the limit process we shrink ΔV toward zero volume around a particular point. We will identify the point by its position in a three-dimensional Cartesian coordinate system, and we'll denote the position by x. The components corresponding to x may be (x, y, z) in a rectangular coordinate system, or (r, θ, z) in a cylindrical coordinate system, or other coordinate names in systems of other flavors. Since the volume ΔV shrinks up to the point x, we say $\rho = \rho(x)$. In more general problems, time may also be involved and $\rho = \rho(x,t)$. The only reservation we need in order to apply all of these ideas to soil is that, rather than take the limit as $\Delta V \to 0$, we will agree to consider finite volumes and restrain the limiting procedure to $\Delta V \to \Delta V_o$, where ΔV_o is sufficiently large to contain a significant number of soil particles.

The concept of a "permissible volume" ΔV_o becomes quite important especially when dealing with the mechanical testing of soils for the determination of their strength and deformability characteristics. For example, in the triaxial testing of soils such as clay, silt, and sand, the dimensions of a cylindrical sample of the soil can be 75 mm in diameter and 150 mm in length. Clearly, the same sample dimensions will not apply when testing granular materials such as gravel with particle sizes of the order of 50 mm. Here, the sample size must be substantially larger if we are to be assured that the results derived from the experiments conform to our notion of a continuum description.

1.3 Deformations of a continuum

When dealing with continua, one important aspect is the description of its deformations. We will need to be able to precisely describe the *deformations* that a continuum may experience as a result of action from outside forces. The term deformation refers both to the motion of a particular particle in the continuum and to the overall motion of the continuum itself. To be a little more specific, suppose our continuum is a *body*, B, which we could illustrate as a generic potato-ish shape in Figure 1.3. Later on we will be much more specific about the configuration of the body, but for the time being the gen-

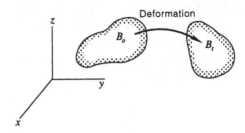

Figure 1.3 The reference configuration B_o and the deformed configuration B_t for a generic elastic body.

eral shape is all we need. We will let the body have a *reference configuration*, and this will be its shape and position when it is at rest, free from load. We will denote the reference configuration by B_o. We can also talk about its *deformed configuration*, after the loads have been applied, and we shall denote this by B_t. The subscript t here refers to time. The configuration B_t gives us the shape and position of the body at time t. We may need to be specific about t in cases where the applied loads vary with time. In Figure 1.3, the reference and deformed configurations are both shown and are linked by a generalized idea of deformation, shown as an arrow linking the two configurations.

How can we precisely describe the deformation? One way to accomplish this is to introduce a vector, called the *displacement vector u*, which joins the position of a particular particle in the reference configuration to its position in the deformed configuration. A typical displacement vector is illustrated in Figure 1.4. If we define a displacement vector for every particle in the body, then the complete set of vectors forms a *vector field*, and we can write

$$u = u(x,t) \qquad (1.1a)$$

to show that u depends both on the position x and the time t.

It is also convenient to use an indicial notation to present the dependent and independent variable encountered in the presentation of basic results. For example, in indicial notation, the position vector is denoted by x_i where the subscript i can take the values 1, 2, 3, and for convenience we can denote $x_1 = x$, $x_2 = y$, $x_3 = z$. Consequently, the displacement vector u can be written as

$$u_i = u_i(x_j,t). \qquad (1.1b)$$

In using the indicial notation we employ the summation convention adopted by Einstein in that if two indices are repeated in an equation, summation is

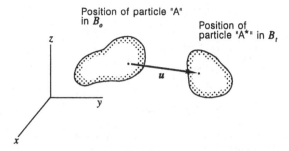

Figure 1.4 The displacement vector *u* joins the position "A" for a particle in the reference configuration to its position "A*" in the deformed configuration.

carried out over the repeated index unless otherwise explicitly stated to the contrary.

It is natural to wonder about the position *x* in eq. (1.1). Is *x* the position of the particle in the reference configuration, or is it the position taken by the particle in the deformed configuration? In a general development of continuum mechanics, *x* in eq. (1.1) would be the position in the reference configuration, but, in the linear theory of elasticity, we assume the *deformations are small*, so small that the distinction between the positions in the two configurations will not be important. That does not mean *u* itself is unimportant, only that whether we use the reference configuration or the deformed configuration positions as an independent variable is not important.

If *u* is known for a certain deformation, then we really have everything we need to know. The vector field *u* gives us the deformation of the body as a whole, and, if we are interested in a specific particle, then $u(x,t)$ evaluated for the appropriate *x* gives us the displacement of the particle as a function of time. Later on, when we are dealing with foundations, *u* will tell us how much the foundation moves. The vertical component of *u*, evaluated at a point immediately beneath the foundation, will give the *settlement* of that point.

1.4 Deformation and strain

We intuitively feel that deformations may lead to *strains* within a body. Before we can consider strains, however, we need to realize that some deformations will not cause strains. These are deformations called *rigid body deformations*, and they consist of either *rigid translations* or *rigid rotations*. A rigid translation is any deformation that does not depend on *x*. Thus, if *u* for every *x* is the same, the body must be undergoing a rigid translation. Rigid rotations rotate the body about a fixed axis. We can be more specific about

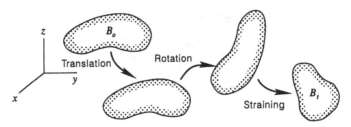

Figure 1.5 Any deformation can be decomposed into a sequence of rigid translation followed by rigid rotation followed by straining.

rotations in a moment. For the time being, the important thing to realize is that *any* deformation can be broken down into, at most, a rigid translation followed by a rigid rotation followed by straining. This idea is shown schematically in Figure 1.5.

The difference between straining and the two rigid motions is this: Strains result in changes in length or shape within the body; rigid motions do not. Suppose we consider two particles within the body that are quite close together in the reference configuration. Let the line that joins these particles be $d\boldsymbol{x}$. We can think of $d\boldsymbol{x}$ as a small filament of the material, and we can examine what happens to this filament during the deformation. If there is any rigid translation, the filament will move, but its orientation will remain unchanged and its length will not be changed either. If there is also rigid rotation, the filament will change its orientation, but its length will still be unchanged. If the filament is stretched or compressed in length, then the body is undergoing straining. We call the change in length divided by the original length the *extensional strain*.

$$\text{extensional strain} = \frac{\text{change in filament length}}{\text{original filament length}}$$

Changes in shape also result in strains. Consider two material filaments $d\boldsymbol{x}_1$ and $d\boldsymbol{x}_2$ which, in the reference configuration, lie at right angles to each other, as in Figure 1.6. If the angle between the two filaments changes during the deformation, then, even if both filaments still have their original length, there has been straining. This is called *shear straining*. The shear strain is defined as the decrease in the angle between the two filaments.

The question we now need to answer is this: How can we go about separating the strains from the rigid motions? Suppose we were given a specified displacement field $\boldsymbol{u}(\boldsymbol{x},t)$. We are aware $\boldsymbol{u}$ may consist of a rigid translation,

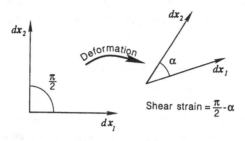

Figure 1.6 Shear strain is defined as the decrease in an initially right angle between two material filaments.

a rigid rotation, and straining. Only the strains will result in stresses within the body. If we want to characterize the stresses, we need to find the strains, and that means eliminating the rigid motions. The first step is to look at how u varies in the neighborhood of a point; i.e., we look at the partial derivatives of u rather than u itself. There will be nine partial derivatives in all. In a rectangular coordinate system each of the components u_x, u_y, and u_z will have three derivatives, one in each coordinate direction x, y, z. If the deformation were *only* a rigid translation, then u would be independent of the variables, and all the partial derivatives would be zero. So by considering the derivatives we have concentrated on rigid rotations and strains and eliminated rigid translations.

Next we need to separate rotations and strains. We can arrange the nine partial derivatives of u into a matrix called the *displacement gradient matrix*, ∇u.

$$\nabla u = \begin{bmatrix} \dfrac{\partial u_x}{\partial x} & \dfrac{\partial u_x}{\partial y} & \dfrac{\partial u_x}{\partial z} \\[2mm] \dfrac{\partial u_y}{\partial x} & \dfrac{\partial u_y}{\partial y} & \dfrac{\partial u_y}{\partial z} \\[2mm] \dfrac{\partial u_z}{\partial x} & \dfrac{\partial u_z}{\partial y} & \dfrac{\partial u_z}{\partial z} \end{bmatrix} \tag{1.2}$$

The matrix for ∇u would look a little different if we were using a cylindrical or spherical coordinate system, but that need not worry us here. The next step is to decompose ∇u into two matrices, one symmetric and one skew-symmetric. The symmetric matrix is called the *strain matrix*, ϵ, and is defined by

$$\epsilon = \frac{1}{2}[\nabla u + (\nabla u)^T]. \tag{1.3}$$

Here the superposed T denotes the transpose of the matrix. The skew-symmetric matrix is called the *rotation matrix*, Ω, defined by

$$\Omega = \frac{1}{2}[\nabla u - (\nabla u)^T]. \tag{1.4}$$

Note that if we add ϵ and Ω we get ∇u. This decomposition into symmetric and skew-symmetric parts is unique and can be accomplished with any square matrix. Just as their names imply, ϵ will correctly account for the strains in the body and Ω will account for the rigid rotations.

We will write the components of the strain matrix like this

$$\epsilon = \begin{bmatrix} \epsilon_{xx} & \epsilon_{xy} & \epsilon_{xz} \\ \epsilon_{yx} & \epsilon_{yy} & \epsilon_{yz} \\ \epsilon_{zx} & \epsilon_{zy} & \epsilon_{zz} \end{bmatrix} \tag{1.5}$$

and bear in mind that symmetry of ϵ implies $\epsilon_{xy} = \epsilon_{yx}$, $\epsilon_{xz} = \epsilon_{zx}$, and $\epsilon_{yz} = \epsilon_{zy}$. The diagonal components of ϵ, ϵ_{xx}, ϵ_{yy}, and ϵ_{zz}, are called the extensional strains. The off-diagonal components, ϵ_{xy}, etc., are called the shear strains. In terms of the partial derivatives of u, we have

$$\epsilon_{xx} = \frac{\partial u_x}{\partial x}, \quad \epsilon_{yy} = \frac{\partial u_y}{\partial y}, \quad \epsilon_{zz} = \frac{\partial u_z}{\partial z} \tag{1.6}$$

$$\left. \begin{array}{c} \epsilon_{xy} = \epsilon_{yx} = \frac{1}{2}\left(\frac{\partial u_x}{\partial y} + \frac{\partial u_y}{\partial x} \right), \quad \epsilon_{xz} = \epsilon_{zx} = \frac{1}{2}\left(\frac{\partial u_x}{\partial z} + \frac{\partial u_z}{\partial x} \right) \\ \\ \epsilon_{yz} = \epsilon_{zy} = \frac{1}{2}\left(\frac{\partial u_y}{\partial z} + \frac{\partial u_z}{\partial y} \right) \end{array} \right\} \tag{1.7}$$

We will see how these names come about by considering two simple examples.

The first example is shown in Figure 1.7. We have a material filament of length dx initially, aligned with the x-axis. (We can consider dx as just the scalar length of the filament whose vector description is $d\mathbf{x} = dx\, \hat{\imath}$, where $\hat{\imath}$ is the unit base vector in the x-direction.) This filament joins points A and B in the reference configuration. In the deformed configuration A has moved to A' and B to B'. The filament length has now changed to $dx + (\partial u_x/\partial x)dx$,

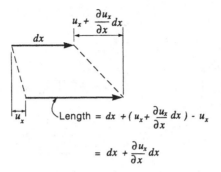

Figure 1.7 An example of extensional strain.

as shown in the figure. Then, according to our definition of extensional strain:

$$\text{extensional strain} = \frac{\text{change in length of filament}}{\text{original length of filament}}$$

$$= \frac{\left(dx + \dfrac{\partial u_x}{\partial x}\,dx\right) - dx}{dx}$$

$$= \frac{\partial u_x}{\partial x}$$

$$= \epsilon_{xx}$$

which shows we were correct to refer to ϵ_{xx} as an extensional strain. It is, of course, the strain in the x-direction. In a similar way we could show ϵ_{yy} and ϵ_{zz} are the extensional strains in the y- and z-coordinate directions respectively.

The second example is shown in Figure 1.8, where two filaments of lengths dx and dy are initially aligned with the x- and y-coordinate axes. Initially the angle between the two filaments is 90°. In the deformed configuration the angle will have changed, as illustrated in Figure 1.8. Earlier we defined the shear strain as the decrease in the initially right angle between the two filaments:

$$\text{shear strain} = \text{decrease in angle between filaments}$$

$$= \alpha + \beta$$

$$\approx \tan\alpha + \tan\beta$$

$$= \frac{\partial u_x}{\partial y} + \frac{\partial u_y}{\partial x}.$$

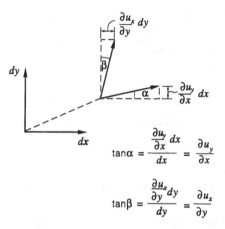

$$\tan\alpha = \frac{\frac{\partial u_y}{\partial x}dx}{dx} = \frac{\partial u_y}{\partial x}$$

$$\tan\beta = \frac{\frac{\partial u_x}{\partial y}dy}{dy} = \frac{\partial u_x}{\partial y}$$

Figure 1.8 An example of shear strain.

Note that we have an implied assumption that α and β are *small* angles and hence $\alpha \approx \tan\alpha$ and $\beta \approx \tan\beta$. If we compare this result with eq. (1.7) we see that ϵ_{xy} is just *one-half* the shear strain. The reason for it being one-half is to make the strain matrix ϵ behave properly under coordinate transformation. In a similar way we could show that ϵ_{xz} and ϵ_{yz} are one-half the angle decreases for filaments aligned with the x- and z-direction and the y- and z-direction.

It is important to bear in mind here that these definitions depend on the provision that the deformations, and particularly the components of the deformation gradient matrix ∇u, are small. Only when we have small deformations will eqs. (1.6) and (1.7) be good representations for extensional and shear strains.

Finally we need to consider the rotation matrix Ω. In component form we have

$$\Omega = \begin{bmatrix} 0 & \Omega_{xy} & \Omega_{xz} \\ \Omega_{yx} & 0 & \Omega_{yz} \\ \Omega_{zx} & \Omega_{zy} & 0 \end{bmatrix} \tag{1.8}$$

where

$$\left. \begin{array}{c} \Omega_{xy} = -\Omega_{yx} = \dfrac{1}{2}\left(\dfrac{\partial u_x}{\partial y} - \dfrac{\partial u_y}{\partial x}\right), \quad \Omega_{xz} = -\Omega_{zx} = \dfrac{1}{2}\left(\dfrac{\partial u_x}{\partial z} - \dfrac{\partial u_z}{\partial x}\right) \\[3mm] \Omega_{yz} = -\Omega_{zy} = \dfrac{1}{2}\left(\dfrac{\partial u_y}{\partial z} - \dfrac{\partial u_z}{\partial y}\right) \end{array} \right\} . \tag{1.9}$$

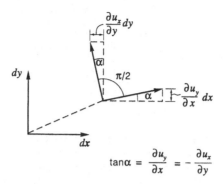

$$\tan\alpha = \frac{\partial u_y}{\partial x} = -\frac{\partial u_x}{\partial y}$$

Figure 1.9 An example of rigid rotation.

Because of its skew symmetry, Ω has only three independent nonzero components. A simple example to illustrate the meaning of these components is shown in Figure 1.9. We have two filaments dx and dy initially aligned with the x- and y-directions. If the body undergoes a rigid rotation, the filaments will be rotated but the 90° angle between them will remain unchanged. The angle of rotation α is, for small rotations, equal to $\partial u_y/\partial x$ and equal to $-\partial u_x/\partial y$. Comparing with eq. (1.9) we see that

$$\Omega_{yx} = \frac{1}{2}\left(\frac{\partial u_y}{\partial x} - \frac{\partial u_x}{\partial y}\right) = \frac{1}{2}(\alpha - (-\alpha)) = \alpha.$$

Had the rotation been clockwise rather than counterclockwise, the signs would be reversed and we would have found Ω_{xy} equal to α. Thus, Ω_{xy} and Ω_{yx} represent the rigid rotation about the z-axis that the body has sustained. Similarly, $\Omega_{xz} = -\Omega_{zx}$ gives the (clockwise) rotation about the y-axis, and $\Omega_{yz} = -\Omega_{zy}$ gives the rotation about the x-axis. Since there are only three degrees of freedom for rotation, Ω contains a complete description of the rigid rotations that the body has suffered. Our earlier provisions concerning small deformations apply here too.

1.5 Volumetric strain

Changes in length within the body may result in changes of volume. We will define the volumetric strain:

$$\text{volumetric strain} = \frac{\text{change of volume}}{\text{original volume}}$$

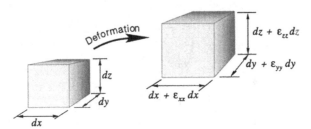

Figure 1.10 Volumetric strain e is the change in volume of a material element divided by its original volume.

and denote this by e. To see how e is related to ϵ, consider the elemental cube of material shown in Figure 1.10. In the reference configuration the lengths of the sides of the cube are dx, dy, and dz. The volume in the reference configuration is $dx \, dy \, dz$. In the deformed configuration the lengths of the sides will be changed as shown in the figure. The volume in the deformed configuration will be:

$$\text{deformed volume} = (dx + \epsilon_{xx}dx)(dy + \epsilon_{yy}dy)(dz + \epsilon_{zz}dz)$$

$$= dxdydz(1 + \epsilon_{xx})(1 + \epsilon_{yy})(1 + \epsilon_{zz}).$$

The volumetric strain is therefore

$$e = \frac{\text{deformed volume} - \text{original volume}}{\text{original volume}}$$

$$= (1 + \epsilon_{xx})(1 + \epsilon_{yy})(1 + \epsilon_{zz}) - 1$$

$$= \epsilon_{xx} + \epsilon_{yy} + \epsilon_{zz} + \epsilon_{xx}\epsilon_{yy} + \epsilon_{xx}\epsilon_{zz} + \epsilon_{yy}\epsilon_{zz} + \epsilon_{xx}\epsilon_{yy}\epsilon_{zz}.$$

Now if we recall that all the components of ∇u are small, we see that all of the strains must also be small, and the products of two or more strains, such as $\epsilon_{xx}\epsilon_{yy}$, must be *very* small. This suggests we can neglect all the products of strains and have

$$e = \epsilon_{xx} + \epsilon_{yy} + \epsilon_{zz}. \tag{1.10}$$

The volumetric strain is simply the sum of the extensional strains, so long as the strains are small. Often the sum of the diagonal components of a matrix is called the *trace* of the matrix, abbreviated *tr*. In that notation we have $e = tr\epsilon$.

1.6 Compatibility of strains

There are certain mathematical relationships that the components of the strain matrix must obey. They are called *compatibility conditions*, and physically they express the idea that, after deformation, the body is still a continuum in the sense that it has not suddenly developed any cracks or gaps or any points at which the material overlaps itself. Particles that were originally close together will still be close in the deformed configuration and, moreover, the relative position of several particles will remain unchanged, as illustrated in Figure 1.11.

Mathematically we can look on compatibility this way. If we were given a realistic displacement field $u(x,t)$, it would be easy to use eqs. (1.6) and (1.7) to find the strain components. But what if we were first given the strains and then asked to find the displacements? This is a more interesting task because now we have *six* equations [(1.6) and (1.7)] to solve for only *three* unknown components of displacement. Whether or not we can integrate eqs. (1.6) and (1.7) to give consistent displacements turns out to be exactly equivalent to the physical problem of ensuring no gaps or overlaps have developed.

The whole problem was solved by the great French mathematician Barre de Saint-Venant in 1860. Saint-Venant showed that the strain components must satisfy these six differential equations.

$$\frac{\partial^2 \epsilon_{xx}}{\partial y^2} + \frac{\partial^2 \epsilon_{yy}}{\partial x^2} = 2\frac{\partial^2 \epsilon_{xy}}{\partial x \partial y}$$

$$\frac{\partial^2 \epsilon_{yy}}{\partial z^2} + \frac{\partial^2 \epsilon_{zz}}{\partial y^2} = 2\frac{\partial^2 \epsilon_{yz}}{\partial y \partial z}$$

$$\frac{\partial^2 \epsilon_{zz}}{\partial x^2} + \frac{\partial^2 \epsilon_{xx}}{\partial z^2} = 2\frac{\partial^2 \epsilon_{xz}}{\partial x \partial z}$$

$$\frac{\partial^2 \epsilon_{xx}}{\partial y \partial z} = -\frac{\partial^2 \epsilon_{yz}}{\partial x^2} + \frac{\partial^2 \epsilon_{zx}}{\partial x \partial y} + \frac{\partial^2 \epsilon_{xy}}{\partial x \partial z} \qquad (1.11)$$

$$\frac{\partial^2 \epsilon_{yy}}{\partial z \partial x} = -\frac{\partial^2 \epsilon_{zx}}{\partial y^2} + \frac{\partial^2 \epsilon_{xy}}{\partial y \partial z} + \frac{\partial^2 \epsilon_{yz}}{\partial y \partial x}$$

$$\frac{\partial^2 \epsilon_{zz}}{\partial x \partial y} = -\frac{\partial^2 \epsilon_{xy}}{\partial z^2} + \frac{\partial^2 \epsilon_{yz}}{\partial z \partial x} + \frac{\partial^2 \epsilon_{zx}}{\partial z \partial y}$$

These six *compatibility equations* play an important role in the solution of elasticity problems, but we will not make much use of them, since we will be more concerned with applying known solutions than finding new ones. A detailed derivation of the compatibility equations is given in Appendix A.

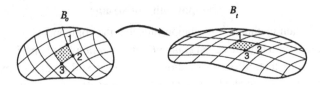

Figure 1.11 The concept of strain compatibility. Points 1, 2, and 3 are found in the deformed configuration in the same relation to one another as they originally were in the reference configuration.

1.7 Stress

If the body is deformed and strains occur, then stresses will be developed within the material. We'll explore the relationship between stress and strain presently, but first we need to be precise about the concept of stress itself. We think intuitively of stress as the ratio of force to area on some surface in the body. For example, consider the rope illustrated in Figure 1.12. If we make a horizontal cut in the rope we can define the *traction vector*, T, as the force in the rope (which is a vector quantity) divided by the cross-section area of the rope:

$$\text{traction vector} = T = \frac{\text{force vector}}{\text{area}}.$$

This idea of force per unit area is intuitive and elementary and well-known to anyone remotely interested in mechanics. But the situation is not as simple as this. What if we were to cut the rope with another surface? Would the traction vector be the same? What if the surface were parallel to the rope instead of perpendicular? These are questions that vexed the mathematicians and mechanicians of the 18th century. It is clear the traction vector will depend on the particular surface we consider. The problem is how to completely describe the state of stress at a given point in the body. We can pass an infinitude of surfaces through the point. In general there may be a different traction on every surface. If we want to characterize the stress state, must we deal with all these tractions?

One such early attempt, describing force exerted by fluid on fluid, and attributed to Archimedes (A.D. 250), is documented by Truesdell (1961). "Let it be supposed that a fluid is of such a character that its parts lying evenly and being continuous, that part which is thrust the less is driven along by that which is thrust the more; and that each of its parts is thrust by the fluid which is above it in a perpendicular direction if the fluid be sunk in anything or

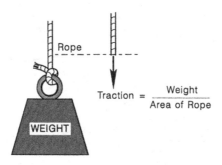

Figure 1.12 The intuitive idea of traction and stress.

pressed together by anything else." As Truesdell remarks, "What this means is not clear."

The answers to all these questions were provided by Augustin Cauchy, a French mathematician, in 1823. Cauchy showed how to find the traction on any surface from knowledge of the tractions on three specific surfaces. Suppose we return to a generic-shaped body as in Figure 1.13. We can look at a surface inside the body and consider an elemental area of that surface, the area ΔA in the figure. Let ΔF be the force acting on ΔA. (This force is due to the action of the material on the right-hand side of the surface acting on the material on the left-hand side. We could look at the surface from the other side and see the same force caused by the left-hand material acting on the right-hand material. That force would act in the opposite direction to ΔF.) The traction vector is defined by the limit

$$T = \lim_{\Delta A \to 0} \frac{\Delta F}{\Delta A}.$$

Next let $\hat{n}$ be a unit vector that is normal to the surface. Cauchy showed that we can determine T from the product of a matrix (the *stress matrix*) with the vector $\hat{n}$.

$$T = \sigma^T \hat{n} \tag{1.12}$$

Here σ is the stress matrix. It contains all the information needed to find the traction on every surface that passes through the point in question. This innocent-looking equation is very important. Just by knowing the nine components of σ (and only six will be independent) we can completely determine the infinitely many tractions that act on the infinitude of surfaces passing through any point. Cauchy's discovery broke open a log-jam of ideas in

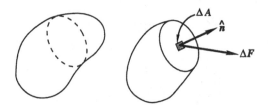

Figure 1.13 If we cut the body in some arbitrary surface, forces from one part of the body acting on the other part will be revealed. These forces are vector quantities.

mechanics. Before, researchers had continually struggled with the description of stress. Eq. (1.12) clarified the situation totally. Cauchy's idea and the derivation of Eq. (1.12) are discussed further in Appendixes B and C.

Of course, we still don't know what the components of $\boldsymbol{\sigma}$ are, or how to find them. Let's write $\boldsymbol{\sigma}$ in component form as

$$\boldsymbol{\sigma} = \begin{bmatrix} \sigma_{xx} & \sigma_{xy} & \sigma_{xz} \\ \sigma_{yx} & \sigma_{yy} & \sigma_{yz} \\ \sigma_{zx} & \sigma_{zy} & \sigma_{zz} \end{bmatrix}. \tag{1.13}$$

To see physically what these components represent, consider a simple example. Suppose the surface ΔA is perpendicular to the x-axis. Then, the unit normal vector $\hat{n}$ will be

$$\hat{n} = \begin{bmatrix} 1 \\ 0 \\ 0 \end{bmatrix}$$

and, from Cauchy's eq. (1.12), the components of the traction vector on this surface will be

$$T = \begin{bmatrix} T_x \\ T_y \\ T_z \end{bmatrix} = \begin{bmatrix} \sigma_{xx} & \sigma_{xy} & \sigma_{xz} \\ \sigma_{yx} & \sigma_{yy} & \sigma_{yz} \\ \sigma_{zx} & \sigma_{zy} & \sigma_{zz} \end{bmatrix}^T \begin{bmatrix} 1 \\ 0 \\ 0 \end{bmatrix} = \begin{bmatrix} \sigma_{xx} \\ \sigma_{xy} \\ \sigma_{xz} \end{bmatrix}.$$

So the stress component σ_{xx} is the component T_x of the traction *that acts on the surface perpendicular to the x-axis.* Similarly σ_{xy} and σ_{xz} are the y- and z-components of the traction T. The top row of the stress matrix consists of the three components of the traction vector that act on the surface perpendicular to the x-axis. With a parallel argument we can show that the second

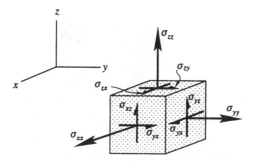

Figure 1.14 The components of the stress matrix.

and third rows of σ are just the components of the tractions that act on surfaces perpendicular to the y- and z-axes. Thus, the only information we need to find σ are the three tractions that act on the three surfaces perpendicular to the three coordinate directions. (See also Appendix C.)

The subscripts on the components of σ now tell us the physical meaning of each component. Consider σ_{xy}, for example. The first subscript, x, tells us this is a component of the traction acting on the surface perpendicular to the x-axis. The second subscript, y, tells us we have the y-component of that traction. All the nine σ components can be summarized by considering an elemental cube of the material with sides perpendicular to the coordinate axes, as shown in Figure 1.14. Just by inspecting this figure we see that the diagonal components σ_{xx}, σ_{yy}, σ_{zz} are the *normal stress* components, or simply the normal stresses, while the off-diagonal components σ_{xy}, etc., act tangential to the surfaces and are the *shear stress* components or shear stresses for short.

Cauchy went on to show, using a limit argument, that σ is also a symmetric matrix. Thus, $\sigma_{xy} = \sigma_{yx}$, etc., and there are only six independent stress components needed to completely characterize the state of stress at any point in the body. The components may change from point to point though, and we can represent this by letting σ be a function of both x and t

$$\sigma = \sigma(x,t).$$

In the theory of elasticity, workers adopt the following sign convention. Consider a surface with unit normal vector pointing in the positive coordinate direction. Any stress component acting on this surface is considered positive if its direction is also in the positive coordinate direction. This is illustrated in Figure 1.15. We can see from this that the normal stress σ_{xx} is positive in *tension*. Unfortunately, the opposite sign convention is usually adopted in

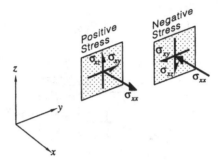

Figure 1.15 Sign conventions for stresses.

geotechnical subjects because compressive normal stresses are more common than tensile ones. This difference is sometimes a small source of confusion in elastic analysis of foundations. We will not use either convention exclusively, but we will be specific about which convention is in use at each stage of the text. For the present chapter we will use the elasticity convention and make tensile stress positive.

Returning again to eq. (1.12), we see in general that the traction vector will act in some arbitrary direction to the surface element. Cauchy's expression gives the components of T in three coordinate directions. In some situations, it may be more interesting to look at the components of T that act normal and tangential to the surface. To find these components, consider Figure 1.16. The vector component perpendicular to the surface is $(T \cdot \hat{n})\hat{n}$. The component tangential to the surface is given by the triple vector product $\hat{n} \times (T \times \hat{n})$. The magnitude of the normal component is called the normal stress acting on the surface. The magnitude of the tangential component is called the shear stress acting on the surface. These are sometimes abbreviated σ and τ. Note that σ and τ are both scalar quantities and both are meaningless unless the surface on which they act is also fully specified. We can express σ and τ as

$$\sigma = T \cdot \hat{n}, \quad \tau = \sqrt{T \cdot T - \sigma^2}. \tag{1.14}$$

A familiar graphical representation for the stress state at a point in terms of τ and σ is the *Mohr diagram*. If we take a certain point in the body and through it pass surfaces of every possible orientation and then graph the points (σ, τ) for each of those surfaces, we find the points all will lie in a well-defined region such as that shown in Figure 1.17. This geometric interpretation of stress was discovered by the German engineer Otto Mohr in 1882. Often, only points on the outermost circle of the diagram may be of interest, but every point of

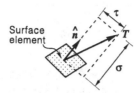

Figure 1.16 The components of traction normal and tangential to the surface.

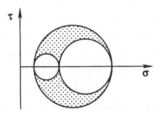

Figure 1.17 A typical Mohr diagram.

the shaded region corresponds to a stress pair (σ, τ), which acts on some surface passing through the point. Different stress states will have different Mohr diagrams with circles of different sizes, but in every case the circle centers will lie on the σ-axis, and there will be three points where $\tau = 0$, corresponding to three surfaces that support no shear stress. These are the principal surfaces we will discuss next.

1.8 Principal stresses

An interesting question to ask at this point is this: Are there any surfaces for which the traction vector T is parallel to the unit normal vector $\hat{n}$? That is, can we find a surface for which

$$T = \lambda\hat{n} \tag{1.15}$$

where λ is a scalar. The answer, of course, is yes. In general, there will be three such surfaces, called *principal surfaces*. The stresses that act on these surfaces will be the *principal stresses*.

To find the principal stresses we can use eq. (1.15) in eq. (1.12) to have

$$\lambda\hat{n} = \sigma^T\hat{n} = \sigma\hat{n},$$

which can be rearranged to become

$$(\boldsymbol{\sigma} - \lambda\mathbf{1})\hat{n} = 0, \tag{1.16}$$

where I denotes the identity matrix. We recognize eq. (1.16) as an eigenvalue problem. A theorem from linear algebra assures us of solutions for (1.16), provided $\boldsymbol{\sigma}$ is symmetric. The eigenvalues are found by solving

$$\det(\boldsymbol{\sigma} - \lambda\mathbf{1}) = 0. \tag{1.17}$$

Expanding the determinant gives a cubic equation for λ. The roots of this equation are the eigenvalues of $\boldsymbol{\sigma}$. We will denote them by σ_1, σ_2, and σ_3, and refer to them as the principal stresses. The corresponding eigenvectors, say $\hat{n}_1$, $\hat{n}_2$, $\hat{n}_3$, define the three principal surfaces. On each surface, the traction is given by the principal stress multiplying the eigenvector. Thus, there is no shear stress on the principal surface. Any surface free of shear stress must be a principal surface. It is convenient to number the principal stresses so that

$$\sigma_1 \geq \sigma_2 \geq \sigma_3.$$

There is another theorem in linear algebra that tells us the eigenvectors will be mutually orthogonal. Thus, the principal surfaces will intersect at right angles. This also means we can find a special coordinate frame, say x^*, y^*, z^*, which aligns (at the point in question) with the eigenvectors. If we express the stress matrix in this coordinate frame it will be

$$\boldsymbol{\sigma} = \begin{bmatrix} \sigma_1 & 0 & 0 \\ 0 & \sigma_2 & 0 \\ 0 & 0 & \sigma_3 \end{bmatrix}. \tag{1.18}$$

It may be convenient to find this special coordinate frame in certain types of problem, but we need to be aware the principal directions may change as we move from one point to another in the body.

1.9 Stress invariants

Suppose we have a body that is subjected to loads and has a nonzero stress matrix at some point of interest. The components of this stress matrix are, as we know, just the components of the three tractions that act on surfaces perpendicular to the three coordinate axes, x, y, and z. Because of this, we see that the components of σ are dependent on our choice of coordinate system. If we had chosen a different coordinate system, say x^*, y^*, z^*, rotated relative to x, y, z, then the stress components would all be different. But, even though the components would be different, there are certain combinations of components that would be the same in either coordinate system. These are called *stress invariants* (invariant implying something that does not change even when the coordinate frame does).

One invariant is the trace of σ or the sum of its diagonal components

$$\text{1st Stress Invariant} = I_1 = tr\sigma = \sigma_{xx} + \sigma_{yy} + \sigma_{zz}. \quad (1.19)$$

This quantity will always be the same regardless of how we orient the x, y, and z axes. This is a fact that can be quite useful. Basic laws of mechanics should not depend on what choice of coordinates we make at the beginning of the problem.

Since we called I_1 the first invariant, it is safe to assume there will be some others. In fact there are infinitely many others since products of I_1 and powers of I_1 are also invariants, *but* there are only two others that are independent of I_1. They are:

$$\text{2nd Stress Invariant} = I_2 = \frac{1}{2}[(tr\sigma)^2 - tr\sigma^2] \quad (1.20)$$

and

$$\text{3rd Stress Invariant} = I_3 = \det\sigma. \quad (1.21)$$

Here $tr\sigma^2$ denotes the trace of σ multiplied by itself. It is that term that makes I_2 independent of I_1 since the other term $(tr\sigma)^2$ is just I_1^2.

The three invariants I_1, I_2, and I_3 are called the *principal invariants* of the stress matrix. They appear in another context with regard to the principal stresses. When we expand the determinant in eq. (1.17), we obtain the characteristic cubic equation in λ, and the coefficients of that equation are the principal invariants. That is, (1.17) is equivalent to

$$-\lambda^3 + I_1\lambda^2 - I_2\lambda + I_3 = 0. \quad (1.22)$$

The roots of this equation are σ_1, σ_2, and σ_3, of course. We can also show that

$$I_1 = \sigma_1 + \sigma_2 + \sigma_3$$

$$I_2 = \sigma_1\sigma_2 + \sigma_2\sigma_3 + \sigma_3\sigma_1$$

$$I_3 = \sigma_1\sigma_2\sigma_3.$$

The easy way to arrive at these equations is to assume the coordinate system aligns with the principal directions so that σ has the diagonal form in eq. (1.18).

All of the ideas we have introduced in the last few pages can be applied to strains as well as stresses. That is, we can define the principal strains as the eigenvalues of the strain matrix. There is an equivalent Mohr diagram for strains, and there are three principal strain invariants just as there are three stress invariants. These strain features are not as interesting for us as their stress counterparts, and we won't need to refer to them any further.

1.10 Equilibrium equations

There are three equations that involve partial derivatives of the stress components and express the concept of static equilibrium in the body. Before writing them down we must first discuss the concept of a *body force*. In a general way, we can classify all forces into two categories: contact forces and body forces. Contact forces are forces associated with surfaces, and they lead to traction vectors such as we discovered in Section 1.6. Body forces are forces associated not with surfaces but with *volumes* within the body. Two familiar types of body forces are gravity forces and magnetic forces. All body forces result from influences outside the body and are assumed to be fully specified for any problem. That is, a body force is *never* an unknown quantity in any equation where it appears. Like any force, body forces will be vector quantities. We will represent the body force by f, and we will agree to let f be the applied body force *per unit volume* of the material. (In some developments f is taken to be the force per unit mass rather than force per unit volume. For those developments we would multiply f by the mass density ρ to get the body force per unit volume.)

Now we can write down the three equations of equilibrium. They must hold at every point in the body, so long as it remains in static equilibrium

$$\frac{\partial \sigma_{xx}}{\partial x} + \frac{\partial \sigma_{xy}}{\partial y} + \frac{\partial \sigma_{xz}}{\partial z} + f_x = 0$$

$$\frac{\partial \sigma_{yx}}{\partial x} + \frac{\partial \sigma_{yy}}{\partial y} + \frac{\partial \sigma_{yz}}{\partial z} + f_y = 0 \qquad (1.23)$$

$$\frac{\partial \sigma_{zx}}{\partial x} + \frac{\partial \sigma_{zy}}{\partial y} + \frac{\partial \sigma_{zz}}{\partial z} + f_z = 0.$$

Here f_x, f_y, and f_z are the components of f, and we have assumed a rectangular coordinate frame is in use. In cylindrical or spherical coordinates, these equations would appear somewhat different.

These three equations each express equilibrium of forces in one of the three coordinate directions. To see how they are obtained, consider the elemental material cube shown in Figure 1.18. The stress components that act on the cube faces perpendicular to the x- and y-directions are shown on the figure, as well as the body force components f_x and f_y. Considering only the forces visible in the figure (some forces are missing) we can write an expression for force equilibrium in the x-direction.

$$\left(\sigma_{xx} + \frac{\partial \sigma_{xx}}{\partial x}\, dx\right)dydz - (\sigma_{xx})dydz + \left(\sigma_{xy} + \frac{\partial \sigma_{xy}}{\partial y}\, dy\right)dxdz$$
$$- (\sigma_{xy})dxdz + f_x dxdydz = 0$$

In this equation, $dx\,dy\,dz$ is the volume of the cube and $f_x\, dx\, dy\, dz$ is the total body force in the x-direction. The quantities $dy\,dz$ and $dx\,dz$ are the areas of the cube faces perpendicular to the x- and y-axes. If we combine the various terms and then divide by $dx\,dy\,dz$, we obtain

$$\frac{\partial \sigma_{xx}}{\partial x} + \frac{\partial \sigma_{xy}}{\partial y} + f_x = 0.$$

This equation is similar to the first of eqs. (1.23), except for the term involving σ_{xz}. But that is precisely the term we ignored by considering only the forces visible on Figure 1.18. So the first of eqs. (1.23) represents force equilibrium in the x-direction. The second and third equations correspond to the y- and z-directions, respectively.

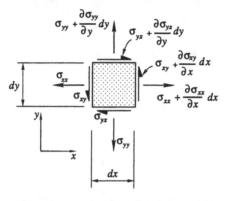

Figure 1.18 A two-dimensional stress state.

1.11 Formulation of problems

To conclude this chapter we will introduce some ideas about how problems in classical elasticity are formulated and solved.

Almost all elasticity problems are formulated as boundary value problems. The body in the reference configuration is subjected to certain prescribed conditions on its boundary and, in responding to these conditions, it deforms. The prescribed conditions may involve either tractions or displacements or both. *Traction boundary value problems* are problems in which the surface traction vector T is prescribed at every point on the boundary of the body. As an example, consider an elastic sphere floating partly submerged in a fluid, as shown in Figure 1.19. On that part of the sphere boundary below the fluid surface, a hydrostatic stress will act. The traction vector at any point will act normal to the boundary, and its magnitude will depend on the depth within the fluid. Above the fluid surface, the tractions are all zero. Thus, the tractions are fully prescribed at every point on the boundary of the sphere.

Displacement boundary value problems are, as the name suggests, problems where the displacement of every point on the boundary is specified. Physically reasonable examples of displacement boundary value problems are not so easy to coin as are examples of traction boundary value problems, although that does not mean displacement problems are of no interest or worth. One problem of real practical interest is inflation of a cavity in an elastic body. Consider the spherical cavity imbedded inside an elastic body illustrated in Figure 1.20. If the body is quite large in relation to the size of the cavity, then we are justified in assuming that changes in cavity volume will not have any effect on the outer boundary, and we might specify the displacements there

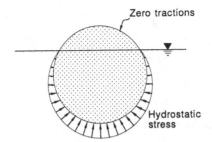

Figure 1.19 An example of a traction boundary value problem.

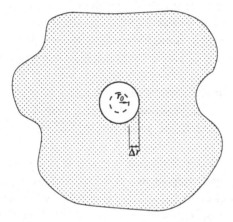

Figure 1.20 An example of a displacement boundary value problem.

must remain zero. Another, probably easier, specification would be to assume the outer boundaries extend so far from the cavity that they are located effectively at infinity and then specify that the displacements must vanish as the radial distance from the cavity approaches infinity. Now we are left with only the boundary of the cavity itself, and on this boundary we specify a constant radial displacement Δr. That displacement might be caused by (in some magical way) increasing the pressure in a fluid that fills the cavity. One might then object and say that if the pressure were known, why not pose the problem as a traction boundary problem, and that might be done as well; but the point here is to illustrate how a displacement problem might arise. Cavity expansion problems such as this play an important role in certain areas of geotechnical engineering.

Mixed boundary value problems are problems in which a part or parts of the boundary have tractions specified, while the remainder of the boundary has displacements specified. A familiar example is the fixed-end beam illus-

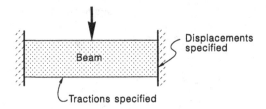

Figure 1.21 An example of a mixed boundary value problem.

trated in Figure 1.21. Tractions are specified on all lateral surfaces of the beam, but zero displacements are specified on both ends.

Mixed-mixed boundary value problems involve boundary conditions that are neither "fish nor fowl." In these problems, at some parts of the boundary, we find a combination of both tractions and displacements specified. An example is the smooth rigid punch indenting an elastic plate shown in Figure 1.22. The bottom surface of the plate rests on a rigid base and displacements are specified zero. The upper surface of the plate, outside the punch, has zero tractions specified. Beneath the punch we find the mixed-mixed conditions. The *vertical* component of displacement will be specified, equal to the punch indentation. But if the face of the punch is smooth so that lateral slip can occur, then the *horizontal* components of the traction must be specified as zero. In the other types of problems we have always had either three components of the traction vector specified or three components of the displacement vector specified. In mixed-mixed problems, either one traction and two displacement components are specified, or one displacement and two tractions components are specified (as for the smooth punch).

Note that in traction boundary value problems only some of the stress components will be specified at the boundary. To illustrate, consider a typical problem in foundation engineering, a uniform circular load applied on the surface of an elastic half-space. The half-space represents the ground and the circular load might be an oil storage tank with a flexible foundation. The problem is shown in Figure 1.23. On the surface of the half-space we have tractions specified, either zero or, beneath the storage tank, vertical tractions of magnitude p_o. What do these specified tractions tell us about the stress components at the boundary $z = 0$? To answer that question we turn to Cauchy's principle, eq. (1.12). The unit normal vector to the boundary and the specified traction vector beneath the tank are given by

$$\hat{n} = \begin{bmatrix} 0 \\ 0 \\ 1 \end{bmatrix}, \quad T = \begin{bmatrix} 0 \\ 0 \\ p_o \end{bmatrix}.$$

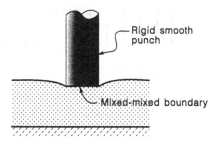

Figure 1.22 An example of a mixed-mixed boundary value problem.

Thus from (1.12) we have

$$\begin{bmatrix} 0 \\ 0 \\ p_o \end{bmatrix} = \begin{bmatrix} \sigma_{xx} & \sigma_{xy} & \sigma_{xz} \\ \sigma_{yx} & \sigma_{yy} & \sigma_{yz} \\ \sigma_{zx} & \sigma_{zy} & \sigma_{zz} \end{bmatrix}^T \begin{bmatrix} 0 \\ 0 \\ 1 \end{bmatrix}$$

from which we can conclude

$$\sigma_{zx} = \sigma_{zy} = 0, \quad \sigma_{zz} = p_o.$$

So one component of normal stress and two shear stresses are specified on the boundary. The remaining components of stress are undetermined. In other problems, a more complicated condition for the stress components may arise if the boundary is not perpendicular to a coordinate direction or if the specified traction does not act perpendicular to the boundary; but in every case, only three equations result from eq. (1.12), and these can ultimately control only three components of stress.

A problem that has been properly formulated is solved by finding stress, strain, and displacement fields for every point in the body such that (i) the equilibrium equations (1.23) are satisfied everywhere and (ii) the specified boundary conditions are satisfied at all points of the boundary. Finding the solution will generally involve solving one or more differential equations. We will not worry about solution techniques, indeed we do not yet have enough equations to solve even the simplest problem. Another six equations that link the components of the stress and strain matrices are still required. These equations represent a generalized form of Hooke's law, and we will wait to discuss them until Chapter 2. The important point is this: Solutions give the components of stress, strain, and displacement everywhere in the body such that equilibrium and all boundary conditions are fully satisfied. It can be shown, if a solution can be found, that it will be unique in terms of stresses and strains, and the displacements will be unique

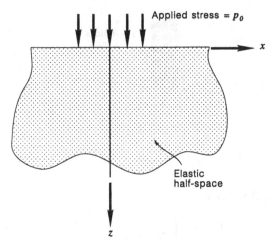

Figure 1.23 An elastic half-space supporting a uniform applied load.

to within an arbitrary rigid motion. By that we mean two different solutions to the same problem cannot exist, except for solutions that differ only by rigid translations and rigid rotations. This fact is discussed further in Appendix D.

One final point to end this chapter: The linear nature of the equations of classical elasticity allows us to use superposition. Therefore, two solutions for the same body but with different boundary conditions can be added together to give the solution for the problem where both sets of boundary conditions are applied simultaneously. This is an extraordinary powerful tool, and we will exploit it continuously in later chapters.

Exercises

1.1 Consider the deformation defined by this displacement vector field

$$u_x = kxy^2, \quad u_y = u_z = 0.$$

Write out the displacement gradient matrix, the strain matrix, and the rotation matrix. Verify that the Saint-Venant compatibility conditions [eqs. (1.11)] are satisfied by this deformation.

1.2 The deformation illustrated below is called simple shear. The displacement vector field is given by

$$u_x = \tan\beta y, \quad u_y = u_z = 0.$$

Write out the displacement gradient matrix, the strain matrix, and the rotation matrix. Verify that the nonzero components of the strain and rotation matrices agree with our definitions for shear strain and rigid rotation. Sketch how the deformed configuration can be obtained by a rigid rotation followed by pure shearing.

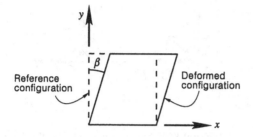

1.3 At some point in an elastic body the stress matrix is given by

$$\sigma = \begin{bmatrix} 20 & 0 & 0 \\ 0 & 10 & 0 \\ 0 & 0 & 5 \end{bmatrix}$$

where the stress components are measured in kPa.

(i) On one graph, plot the family of traction vectors that act on surfaces with the following unit normal vectors

$$\begin{bmatrix} 1 \\ 0 \\ 0 \end{bmatrix}, \quad \begin{bmatrix} \sqrt{3}/2 \\ 1/2 \\ 0 \end{bmatrix}, \quad \begin{bmatrix} 1/\sqrt{2} \\ 1/\sqrt{2} \\ 0 \end{bmatrix}, \quad \begin{bmatrix} 1/2 \\ \sqrt{3}/2 \\ 0 \end{bmatrix}, \quad \begin{bmatrix} 0 \\ 1 \\ 0 \end{bmatrix}.$$

That is, find $T = \sigma^T \hat{n}$ for each $\hat{n}$ and graph the resulting vectors from a common origin.

(ii) Demonstrate that the endpoints of the traction vectors in part (i) lie on an ellipse.

(iii) Generalize your result in part (ii) to prove that the locus of endpoints of all tractions T that act on all surfaces at any point in an elastic body must lie on an ellipsoid. This is called the Lamé stress ellipsoid.

1.4 For the stress matrix given in question 1.3, find the principal stresses and the orientations of the principal planes.

1.5 For the stress matrix given in question 1.3, find the three principal stress invariants. Use these together with the principal stresses found in question 1.4 to verify that the characteristic equation (1.22) is satisfied for all three principal stresses.

1.6 The octahedral plane is defined as the surface whose unit normal vector is equally inclined to the principal stress directions. This corresponds to

the plane ABC below. By considering equilibrium of the wedge OABC, derive expressions (in terms of principal stresses) for the following:
(i) the normal stress on the octahedral plane (σ_{oct});
(ii) the shear stress on the octahedral plane (τ_{oct}).

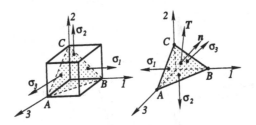

1.7 Let T_1 and T_2 be traction vectors at point x that act on surfaces with unit normals $\hat{n}_1$ and $\hat{n}_2$. Prove that

$$T_1 \cdot \hat{n}_2 = T_2 \cdot \hat{n}_1.$$

What property of the stress matrix is necessary for your proof?

1.8 The stress matrix at some point in an elastic body is known to be

$$\sigma = \begin{bmatrix} 1 & 1 & 0 \\ 1 & -1 & 0 \\ 0 & 0 & 1 \end{bmatrix}.$$

Consider the surface passing through this point whose normal vector is parallel to [1, 1, 2].
(i) Find the components of the traction vector that acts on this surface.
(ii) Find the magnitudes of normal and shear stress that act on this surface.

1.9 A true triaxial testing device has been developed for laboratory testing of cubical samples of soils. In this device, the principal stresses can be independently controlled. A test is conducted using the device to evaluate the strength characteristics of a dry sand. At failure, the test gives the following values for the stress invariants

$$I_1 = 6 \times 10^2 \text{ kPa}$$

$$I_2 = 11 \times 10^4 \text{ (kPa)}^2$$

$$I_3 = 6 \times 10^6 \text{ (kPa)}^3.$$

(i) Calculate the principal stresses at failure.
(ii) What is the greatest shear stress on any surface within the sample at failure?

1.10 A hollow sphere of diameter $2a$ and weight W is pushed downward into a slurry of thixotropic clay used for drilling. The slurry can be treated as a nonviscous fluid. *Without* using Archimedes' Principle derive an expression for the unit weight of the slurry if a force P is required to immerse the sphere just below the slurry surface as noted in the figure.

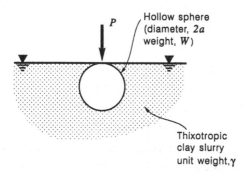

References

The two works mentioned in Section 1.1 are:

Cauchy, A. L., "Recherches sur l'Équilibre et le Mouvement Intérieur des Corps Solides ou Fluides, Élastiques ou non Élastiques," *Bull. Soc. Philomath* Vol. 2, 300–304 (1823).

Mohr, O., *Zivilingenieur*, W. Ernst und Sohn, Berlin (1882).

Saint Venant, Barré de A. J. C., "Établissement élémentaire des formules et équations générales de la théorie de l'élasticité des corps solides, Appendix in: Résumé des Leçons des Ponts et Chaussées sur l'Application de la Mécanique première partie, prèmiere section," *De la Résistance des Corps Solides*, par C.-L. M. H. Navier, 3rd Ed. Paris. (1864).

Poulos, H. G., and Davis, E. H., *Elastic Solutions for Soil and Rock Mechanics*, Wiley, New York (1974).

Selvadurai, A. P. S., *Elastic Analysis of Soil-Foundation Interaction, Developments in Geotechnical Engineering*, Vol. 17, Elsevier, Amsterdam (1979).

Truesdell's remarks on the concept of stress can be found in:

Truesdell, C., Stages in the development of the concept of stress, in *Problems of Continuum Mechanics*, Contributions in Honor of the 70th Birthday of Academician N.I. Muskhelishvili, Society for Industrial and Applied Mathematics, Philadelphia, Pennsylvania, 556–564 (1961).

There are many excellent references on the theory of elasticity. Three modern works that develop the theory are:

Atkin, R. J., and Fox, N., *An Introduction to the Theory of Elasticity*, Longman, London (1980).

Barber, J., *Elasticity*, Kluwer Academic Publishers, AA Dordrecht, The Netherlands (1992).

Fung, Y. C., *Foundations of Solid Mechanics*, Prentice-Hall, New Jersey (1965).

Little, R. W., *Elasticity*, Prentice Hall, New Jersey (1973).

Malvern, L. E., *Introduction to the Mechanics of a Continuous Medium*, Prentice-Hall, New Jersey (1969).

2

The elastic constants

2.1 Hooke's law

Hooke's law is the glue that binds the theory of elasticity together. It provides the missing equations that relate the stresses and strains at all points within the body. The form of Hooke's law we will need is a little more complicated than the form Robert Hooke proposed in 1676. Hooke was concerned with springs instead of three-dimensional continuous bodies, and he simply stated that the force needed to elongate the spring was a linear function of the elongation. We will need to generalize his idea and let the strains be linear functions of the stresses. The result is often called the *generalized Hooke's law.*

The generalized Hooke's law is an example of a *constitutive relation.* In continuum mechanics, constitutive relations are equations that relate causes and effects. Strains and stresses are one example of cause and effect, temperature gradient and heat flux are another, temperature and volumetric strain yet another. Some constitutive relations involve time and can describe viscous behavior and creep. All constitutive equations involve one or more parameters, and these parameters will take different values for different materials. For example, the coefficient of thermal expansion is the ratio of volumetric strain to temperature change which we observe when we heat a body and it expands. The coefficient of thermal expansion for steel is 12×10^{-6} °C^{-1}, while the value for mercury is quite different, 0.18×10^{-3} °C^{-1}. When it comes to the generalized Hooke's law, the parameters are called the *elastic constants.* There are five of them, but they are not independent. Only two are needed to fully describe the elastic behavior of an isotropic body. The restriction to isotropic materials here is quite important. Soils are often not isotropic, but, because it is convenient to do so, we usually assume they are. This is an assumption made in many areas of geotechnical engineering, not only in elastic analysis.

Of the five elastic constants, four relate strains to stresses (or vice versa) while the fifth, Poisson's ratio, relates one extensional strain to another. If any two constants are known, the other three may be found from simple algebraic relationships. In general, it will be necessary to know two constants in order to attack most problems in foundation engineering. There is one special case, however, where one constant will suffice. That is where we are concerned with undrained response of a fully saturated soil. That case, often referred to as the end of construction or the immediate stability case, is one of particular interest.

2.2 Young's modulus and Poisson's ratio

The most familiar elastic constants are *Young's modulus* and *Poisson's ratio*. Both are illustrated by the example of an elastic bar subjected to tension, shown in Figure 2.1. This loading is called uniaxial tension. The lateral surfaces of the bar are traction free, and the ends are subjected to a uniform stress σ_{xx} (assuming the x-axis is aligned with the bar). The stress σ_{xx} will cause the bar to elongate, which results in an extensional strain ϵ_{xx}. Observations show (for some materials) that ϵ_{xx} depends linearly on σ_{xx}.

$$\epsilon_{xx} = \frac{\sigma_{xx}}{E} \tag{2.1}$$

Here E is Young's modulus. It has dimensions of stress.

Equation (2.1) is familiar to all students of elementary mechanics or physics. It will be an adequate description of material response for a wide range of materials, including soils, provided the stress does not become too large in comparison with the failure stress. We discussed this provision in regard to soils in Section 1.1. Of course, it would be unwise to subject a soil sample to uniaxial tension as in Figure 2.1. Soils are generally very weak in tension. We would apply compressive stress and the resulting shortening of the sample would lead to negative extensional strain. Equation (2.1) would still hold since the compressive stress would be negative. (We are still using the sign convention that tension is positive.)

Now returning to Figure 2.1, observations also show that as the bar elongates in response to σ_{xx}, there will also be lateral contraction. The bar grows longer and skinnier simultaneously. This lateral contraction leads to more extensional strains, ϵ_{yy} and ϵ_{zz}, in the lateral directions. For an isotropic, linear elastic material, ϵ_{yy} and ϵ_{zz} will be equal and will both be linear functions of ϵ_{xx}.

$$\epsilon_{yy} = \epsilon_{zz} = -\nu\epsilon_{xx} \tag{2.2}$$

Figure 2.1 Elastic bar in uniaxial tension.

Here ν is Poisson's ratio. It is a dimensionless quantity. Because of (2.1) we can also write

$$\epsilon_{yy} = \epsilon_{zz} = -\frac{\nu}{E} \sigma_{xx}. \qquad (2.3)$$

Equations (2.1) and (2.3) now give all the strains in terms of the applied stress σ_{xx}.

We can generalize these results by considering a cube of material subjected to uniform normal stress on all six faces as shown in Figure 2.2. The extensional strain ϵ_{xx} will depend on all three stresses. There will be a contribution from σ_{xx} exactly like eq. (2.1), but there will also be contributions from σ_{yy} and σ_{zz} similar to eq. (2.3). The resulting expression for ϵ_{xx} will be

$$\epsilon_{xx} = \frac{\sigma_{xx}}{E} - \frac{\nu}{E} \sigma_{yy} - \frac{\nu}{E} \sigma_{zz}. \qquad (2.4)$$

Two similar equations for ϵ_{yy} and ϵ_{zz} are easily written down. Rewriting eq. (2.4) and carrying on with ϵ_{yy} and ϵ_{zz}, we have the following three equations for the three extensional strains

$$\epsilon_{xx} = \frac{1}{E}[\sigma_{xx} - \nu(\sigma_{yy} + \sigma_{zz})]$$

$$\epsilon_{yy} = \frac{1}{E}[\sigma_{yy} - \nu(\sigma_{xx} + \sigma_{zz})] \qquad (2.5)$$

$$\epsilon_{zz} = \frac{1}{E}[\sigma_{zz} - \nu(\sigma_{xx} + \sigma_{yy})].$$

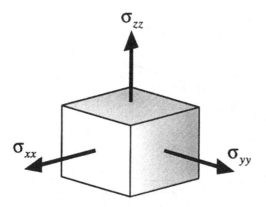

Figure 2.2 Elastic cube subjected to normal tractions.

2.3 Shear modulus

The third elastic constant we want to introduce is the *shear modulus*, G. It has dimensions of stress and it relates the shear stress at a point in the body to the shearing strain that occurs there.

$$\text{shear stress} = G \times \text{shear strain}$$

More specifically, G relates the off-diagonal terms of σ and ϵ as follows

$$\sigma_{xy} = 2G\epsilon_{xy}, \quad \sigma_{yz} = 2G\epsilon_{yz}, \quad \sigma_{zx} = 2G\epsilon_{zx}. \tag{2.6}$$

Recall from Section 1.3 that the off-diagonal components of ϵ: ϵ_{xy}, ..., etc., represent one-half the actual shear strain. This is why the factor of 2 appears in these equations, making (2.6) consistent with the definition above. Also, the symmetry of both σ and ϵ shows us that only the three eqs. (2.6) are required to relate all the six shear stresses to the corresponding six shear strains.

We need to continually bear in mind the assumption that our body is isotropic. If it were not, such simple relationships as (2.6) would not exist.

We can motivate eqs. (2.6) by returning to our example of uniaxial tension in the preceding section. It will also help us to see how the elastic constants are themselves interrelated. Suppose we have the elastic bar in Figure 2.3 subjected to a uniform tensile stress $\sigma_{xx} = p_o$. With the rectangular

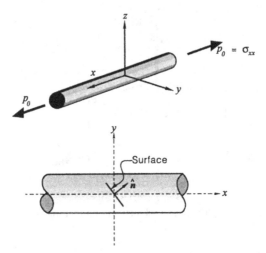

Figure 2.3 Motivation for shear modulus definition.

coordinate system shown, the stress and strain matrices have the following
form

$$\boldsymbol{\sigma} = \begin{bmatrix} p_o & 0 & 0 \\ 0 & 0 & 0 \\ 0 & 0 & 0 \end{bmatrix}, \quad \boldsymbol{\epsilon} = \begin{bmatrix} p_o/E & 0 & 0 \\ 0 & -\nu p_o/E & 0 \\ 0 & 0 & -\nu p_o/E \end{bmatrix}. \quad (2.7)$$

The second of these equations is just eqs. (2.5) recast in matrix form. Now
consider a surface element oriented at an acute angle to the axis of the bar.
Let the unit normal vector to the surface be $\hat{n}$ and let us make things a little
more convenient for ourselves by assuming $\hat{n}$ lies in the x-y plane. Then, in
components

$$\hat{n} = \begin{bmatrix} n_x \\ n_y \\ 0 \end{bmatrix}$$

and the traction vector T, which acts on this surface, follows from eq. (1.12).

$$T = \boldsymbol{\sigma}^T \hat{n} = \begin{bmatrix} p_o n_x \\ 0 \\ 0 \end{bmatrix}$$

The reason we want to look at this surface is because it supports shear stress. The simple geometry of the bar means the surfaces perpendicular to the coordinate directions are all principal surfaces, and they have no shear stress. The oblique surface does, and its magnitude is τ, given by eq. (1.14). We have

$$\sigma = \boldsymbol{T} \cdot \hat{\boldsymbol{n}} = p_o n_x^2$$

and

$$\begin{aligned}
\tau &= \sqrt{\boldsymbol{T} \cdot \boldsymbol{T} - \sigma^2} \\[4pt]
&= \sqrt{(p_o n_x)^2 - p_o^2 n_x^4} \\[4pt]
&= p_o n_x \sqrt{1 - n_x^2} \\[4pt]
&= p_o n_x n_y.
\end{aligned} \tag{2.8}$$

The last equality follows from $\hat{\boldsymbol{n}}$ being a unit vector.

Now find the shear strain that corresponds to τ by considering two material filaments, one parallel to $\hat{\boldsymbol{n}}$ and one perpendicular to $\hat{\boldsymbol{n}}$, before the bar is subjected to load. The angle between the two filaments will initially be 90°. After the load is applied and the bar has elongated, this angle will be reduced by the shearing strain. This is illustrated in Figure 2.4. We will denote the shear strain by γ, and we expect to find that γ is related to τ by the shear modulus:

$$\tau = G\gamma. \tag{2.9}$$

Concentrate for a moment on the upper material filament in Figure 2.4. Let it have length ds_o in the reference configuration, and let dx_o and dy_o be the projected lengths in the x- and y-directions as shown in the figure. Since ds_o is parallel to $\hat{\boldsymbol{n}}$ we have

$$dx_o = n_x ds_o, \quad dy_o = n_y ds_o.$$

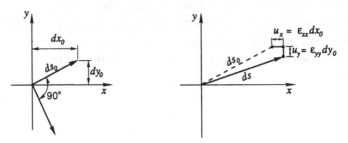

Figure 2.4 Shear strain associated with the surface in Figure 2.3.

The deformation will cause the filament to move as shown on the right-hand side of Figure 2.4. The tip of the filament will move relative to the tail by displacements u_x and u_y given by

$$u_x = \epsilon_{xx}dx_o = \frac{p_o}{E}\,dx_o = \frac{p_o}{E}\,n_x ds_o$$

$$u_y = \epsilon_{yy}dy_o = -\frac{\nu p_o}{E}\,dy_o = -\frac{\nu p_o}{E}\,n_y ds_o.$$

The angle change between the initial filament position and its deformed position is made up of two parts. This is illustrated in Figure 2.5, and the angle change is derived on the figure.

$$\alpha + \beta = \frac{p_o(1+\nu)}{E}\,n_x n_y$$

The shear strain γ will be made up of this angle change plus the change experienced by the lower filament in Figure 2.4. We could analyze the lower filament exactly as we have done for the upper filament, but if we do so we find its angle change is exactly the same as for the upper filament. The reason for this is the simple nature of the deformation we have considered. The shear strain γ is then twice the sum $\alpha + \beta$:

$$\gamma = \frac{2p_o(1+\nu)}{E}\,n_x n_y. \tag{2.10}$$

Now if we compare this result with eq. (2.8) we find

$$\gamma = \frac{2(1+\nu)}{E}\,\tau.$$

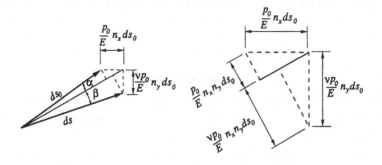

$$\text{Angle change} = \alpha + \beta = \frac{1}{ds_0}\left[\frac{P_0}{E}n_x n_y ds_0 + \frac{\nu P_0}{E}n_x n_y ds_0\right]$$

$$= \frac{P_0(1+\nu)}{E}n_x n_y$$

Figure 2.5 Enlarged view of deformation in Figure 2.4.

That is, γ and τ are linearly related, as we expected, and, comparing this with eq. (2.9),

$$G = \frac{E}{2(1 + \nu)}. \tag{2.11}$$

This equation shows how the shear modulus is fully determined by Young's modulus and Poisson's ratio. In fact, if any two of the three constants E, ν, and G are known, then we can immediately find the third.

Returning to a general state of stress and strain we now have eqs. (2.6) to relate shear strains to shear stresses and eqs. (2.5) to relate extensional strains to normal stresses. All six of these equations are sometimes written in matrix form in a six-dimensional stress-strain vector space:

$$
\begin{bmatrix}
\epsilon_{xx} \\
\epsilon_{yy} \\
\epsilon_{zz} \\
\epsilon_{xy} \\
\epsilon_{yz} \\
\epsilon_{zx}
\end{bmatrix}
=
\begin{bmatrix}
1/E & -\nu/E & -\nu/E & 0 & 0 & 0 \\
-\nu/E & 1/E & -\nu/E & 0 & 0 & 0 \\
-\nu/E & -\nu/E & 1/E & 0 & 0 & 0 \\
0 & 0 & 0 & 1/2G & 0 & 0 \\
0 & 0 & 0 & 0 & 1/2G & 0 \\
0 & 0 & 0 & 0 & 0 & 1/2G
\end{bmatrix}
\begin{bmatrix}
\sigma_{xx} \\
\sigma_{yy} \\
\sigma_{zz} \\
\sigma_{xy} \\
\sigma_{yz} \\
\sigma_{zx}
\end{bmatrix}
\tag{2.12}
$$

This equation, or the corresponding six equations (2.5) and (2.6), are referred to as generalized Hooke's law. They supply the final pieces of information necessary to solve elasticity problems by fully relating the two matrices σ and ϵ.

2.4 Bulk modulus and Lamé constants

There are still two more commonly used elastic constants, the *bulk modulus* and a *Lamé constant*. They are related to E, ν, and G, but have different descriptions.

The bulk modulus, denoted K, relates the sum of the normal stress, $\sigma_{xx} + \sigma_{yy} + \sigma_{zz}$, to the volumetric strain e. We can illustrate it by considering a simple example. Suppose we have a body subjected to a *hydrostatic* or isotropic state of stress. By this we mean the traction vector acts normal to every part of the boundary of the body. The cylinder shown in Figure 2.6 supports a uniform state of hydrostatic stress, p_o. The stress matrix has the simple form

$$\sigma = \begin{bmatrix} p_o & 0 & 0 \\ 0 & p_o & 0 \\ 0 & 0 & p_o \end{bmatrix}.$$

The corresponding extensional strains follow from eqs. (2.5)

$$\epsilon_{xx} = \frac{1}{E}[p_o - \nu(p_o + p_o)] = \frac{p_o(1 - 2\nu)}{E}$$

and ϵ_{yy} and ϵ_{zz} are the same as ϵ_{xx}. Then the volumetric strain is given by eq. (1.10)

$$e = \epsilon_{xx} + \epsilon_{yy} + \epsilon_{zz}$$
$$= \frac{3p_o(1 - 2\nu)}{E}. \tag{2.13}$$

The bulk modulus is defined by

$$p_o = Ke. \tag{2.14}$$

Comparing this with eq. (2.13) we see that

$$K = \frac{E}{3(1 - 2\nu)}. \tag{2.15}$$

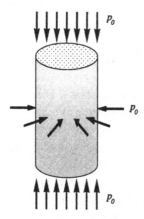

Figure 2.6 Cylindrical body subjected to hydrostatic compression.

For more general stress states, where σ_{xx}, σ_{yy}, and σ_{zz} are all different, the definition (2.14) is generalized to become

$$\frac{1}{3} I_1 = \frac{1}{3} tr\boldsymbol{\sigma} = Ke. \qquad (2.16)$$

Here the first stress invariant I_1 has been used to represent the sum of the normal stresses. The quantity $(^1/_3)I_1$ is commonly called the *mean stress*. Equation (2.16) shows the mean stress and volumetric strain are related by K, the bulk modulus.

The final elastic constant commonly used is the so-called Lamé constant, λ. It has dimensions of stress and is named after the French mechanician Gabriel Lamé, who made many contributions to the theory of elasticity in the early nineteenth century. Unlike the other four elastic constants, λ is not based on a simple physical example. Instead, it arises more as a convenience in writing the normal stresses in terms of the normal strains. We can rearrange eqs. (2.5) and use eq. (2.11) to ultimately find

$$\sigma_{xx} = \lambda e + 2G\epsilon_{xx}$$
$$\sigma_{yy} = \lambda e + 2G\epsilon_{yy} \qquad (2.17)$$
$$\sigma_{zz} = \lambda e + 2G\epsilon_{zz}$$

where

$$\lambda = \frac{\nu E}{(1 + \nu)(1 - 2\nu)} \tag{2.18}$$

and this tells us how the Lamé constant λ is related to E and ν. Equations (2.17) hold exactly the same information as eqs. (2.5), but are inverted to give the normal stresses in terms of the extensional strains. In many cases, it is more convenient to use this form than eqs. (2.5), but there is no new information here. Just a different way of looking at the same thing.

Sometimes the shear modulus G is also referred to as the *second Lamé constant*, and frequently the notation G is changed to μ. Thus, in some books the term shear modulus is not used, but the Lamé constants λ and μ are. Since G and μ are the same thing, this is just a matter of taste or convention. We'll stick with G here and call it the shear modulus.

Using eqs. (2.6) and (2.17) we can now write generalized Hooke's law in another form, linking the matrices σ and ϵ

$$\sigma = \lambda e\mathbf{1} + 2G\epsilon. \tag{2.19}$$

Here $\mathbf{1}$ is the identity matrix. This form of Hooke's law is particularly convenient for general theoretical developments since we are effectively writing down six different equations at once.

2.5 Stress and strain deviators

In the linear theory of elasticity, we can decompose the generalized form of Hooke's law in two completely separate parts, one part dealing with changes in volume and the mean stress, the second part dealing with changes in shape and the part of σ that causes distortion. The first part, dealing with volume changes, can be obtained immediately from eq. (2.19) by taking the trace of both sides of that equation. This gives

$$tr\sigma = I_1 = (3\lambda + 2G)e. \tag{2.20}$$

Here we've used the fact that $tr\mathbf{1} = 3$. If we recall that the mean stress was defined as $(1/3)I_1$, we see that the mean stress and volumetric strain can be related by

$$\frac{1}{3} I_1 = \left(\lambda + \frac{2}{3} G \right) e. \tag{2.21}$$

Comparing this with (2.14) shows that $K = \lambda + (2/3)G$.

It now seems reasonable to expect that there will be a similar simple relationship between some measure of shear stress and a measure of shear strain. What are those measures? They are the *stress and strain deviator matrices*, which we will denote by σ^d and ϵ^d. They are defined by

$$\sigma^d = \sigma - \frac{1}{3}(tr\sigma)\mathbf{1} = \sigma - \frac{1}{3}I_1\mathbf{1} \qquad (2.22)$$

and

$$\epsilon^d = \epsilon - \frac{1}{3}(tr\epsilon)\mathbf{1} = \epsilon - \frac{1}{3}e\mathbf{1}. \qquad (2.23)$$

The form of both these definitions is exactly the same: We take the matrix and subtract one-third its trace from each diagonal component. This is a general definition, which can be usefully applied to other matrices in other contexts.

Now we can use these definitions in eq. (2.19). Replace σ by the equivalent $\sigma^d + (^1/_3)I_1\mathbf{1}$ and replace ϵ by $\epsilon^d + (^1/_3)e\mathbf{1}$. After simplification and using eq. (2.20) we find

$$\sigma^d = 2G\epsilon^d \qquad (2.24)$$

which is the second part of Hooke's law, dealing with shape changes. Together, eqs. (2.21) and (2.24) are completely equivalent to eq. (2.19). This might seem an odd approach since together (2.21) and (2.24) consist of seven equations, whereas (2.19) has only six equations. You might ask if we have inadvertently introduced some new information in (2.21) and (2.24). The explanation lies in the fact that the stress and strain deviators are *traceless*. That is

$$tr\sigma^d = tr\epsilon^d = 0 \qquad (2.25)$$

which we can see immediately by inspecting the definitions (2.22) and (2.23). Equation (2.25) acts as a constraint on eq. (2.24) and in effect tells us that (2.24) can contain only five independent equations. Together, (2.21) and (2.24) comprise only six independent relations, and they are exactly equivalent to (2.19) or to (2.12). Six is the magic number when it comes to Hooke's law. We might also ask: Why should we introduce all these different ways of relating stresses to strains? Surely one form of Hooke's law should suffice? The answer to this is convenience. Different forms will be more convenient in different applications. The mean stress-volumetric strain/stress deviator-strain deviator form will be particularly convenient very soon.

Table 2.1 Relationships between elastic constants

	λ	G	E	ν	K
λ,G			$\dfrac{G(3\lambda + 2G)}{\lambda + G}$	$\dfrac{\lambda}{2(\lambda + G)}$	$\lambda + \dfrac{2}{3}G$
λ,E		$\dfrac{(E - 3\lambda) + \sqrt{(E - 3\lambda)^2 + 8\lambda E}}{4}$		$\dfrac{-(E + \lambda) + \sqrt{(E + \lambda)^2 + 8\lambda^2}}{4\lambda}$	$\dfrac{(E + 3\lambda) + \sqrt{(E + 3\lambda)^2 - 4\lambda E}}{6}$
λ,ν		$\dfrac{\lambda(1 - 2\nu)}{2\nu}$	$\dfrac{\lambda(1 + \nu)(1 - 2\nu)}{\nu}$		$\dfrac{\lambda(1 + \nu)}{3\nu}$
λ,K		$\dfrac{3(K - \lambda)}{2}$	$\dfrac{9K(K - \lambda)}{3K - \lambda}$	$\dfrac{\lambda}{3K - \lambda}$	
G,E	$\dfrac{G(2G - E)}{E - 3G}$			$\dfrac{E - 2G}{2G}$	$\dfrac{GE}{3(3G - E)}$

	λ	G	E	ν	K
G,ν	$\dfrac{2G\nu}{1-2\nu}$		$2G(1+\nu)$		$\dfrac{2G(1+\nu)}{3(1-2\nu)}$
G,K	$K-\dfrac{2}{3}G$		$\dfrac{9KG}{3K+G}$	$\dfrac{3K-2G}{2(3K+G)}$	
E,ν	$\dfrac{\nu E}{(1+\nu)(1-2\nu)}$	$\dfrac{E}{2(1+\nu)}$			$\dfrac{E}{3(1-2\nu)}$
E,K	$\dfrac{3K(3K-E)}{9K-E}$	$\dfrac{3KE}{9K-E}$		$\dfrac{3K-E}{6K}$	
ν,K	$\dfrac{3K\nu}{1+\nu}$	$\dfrac{3K(1-2\nu)}{2(1+\nu)}$	$3K(1-2\nu)$		

Note: To find any particular constant from two others, locate the desired constant in the top row of the table, then the two other constants in the first column. Where the appropriate column and row cross is the desired formula.

2.6 Relationships between elastic constants

Any of the five elastic constants E, v, G, K, and λ can be expressed in terms of any two other constants. We have discovered three of these relationships in eqs. (2.11), (2.15), and (2.18). Other relationships follow from simple algebraic manipulation. Table 2.1 shows all the thirty possible relationships between constants. To find any particular constant from two others, locate the desired constant in the top row of the table, then the two other constants in the first column. Where the appropriate column and row cross is the desired formula.

2.7 Bounds on elastic constants

An interesting question now arises. Can the elastic constants take on any value, or are there constraints they must obey? The answer is that all five of the constants are constrained, at least from below. To see how these bounds come about, we need to introduce the idea of a *stored energy function*. When an elastic body is subjected to specified displacements or tractions on its boundary, the deformations that result imply energy is being stored in the body. That energy can be recovered from the body if the applied loads or specified displacements are later removed. We are only considering the special case when there is no heating or cooling taking place, and the temperature of the body remains constant. An example is stretching a rubber band. If we suddenly release the tractions, the stored energy in the band causes it to resume its at-rest configuration and in the process to fly across the room. Another example is the gradual accumulation of stored energy in the rock mass adjacent to a fault. If the frictional strength of the fault surface is exceeded, a slip occurs and the release of energy results in an earthquake.

To be more specific, consider the familiar example of an elastic bar subjected to uniaxial tension, as shown in Figure 2.7. Let the length of the bar be L and its cross-sectional area be A. If one end of the bar remains stationary, the other end will move a distance $\epsilon_{xx}L$. The force on that end will be $\sigma_{xx}A$. The work done by the force is

$$\text{Work} = \frac{1}{2}(\sigma_{xx}A)(\epsilon_{xx}L) = \frac{1}{2}\sigma_{xx}\epsilon_{xx}AL.$$

The one-half occurs here because the strain increases as the stress increases, and the work is the area beneath the force-elongation curve, as shown in Figure 2.7. No work is done on the lateral surfaces of the bar since they are free from stress.

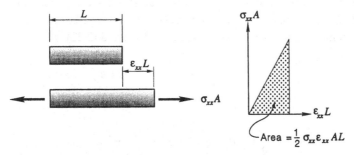

Figure 2.7 Work done on elastic bar in uniaxial tension.

If no energy is lost in the process, the work done on the bar by the applied load is stored as elastic strain energy in the bar. We will define the stored energy function W as the elastic strain energy per unit volume. Since the volume of the bar is AL, we see immediately from above that

$$W = \frac{1}{2}\,\sigma_{xx}\epsilon_{xx}.$$

Now we want to generalize these ideas to the case of an arbitrary deformation of a body of any configuration. The correct definition for the general stored energy function is

$$W = \frac{1}{2}\,tr(\sigma\epsilon). \tag{2.26}$$

Here we have the trace of the product of the stress and strain matrices. W gives the elastic strain energy per unit volume at any point in the body. Clearly W is a scalar and $W = W(x,t)$.

Next, we have a realistic expectation that W will always be greater than or equal to zero. That is because the signs of the stresses and strains will always be the same. Compressive stress should result in shortening, tensile stress in lengthening, clockwise shear stress in clockwise distortion, etc. This seems intuitively obvious, and we accept it as gospel truth. Another way to look at the issue is this: If we increase tractions on the boundary of the body, the energy stored must increase.

In fact, we will assume now that W is a *positive definite function* of its arguments σ and ϵ. That means it will always be greater than or equal to zero and will be zero *only* when the stresses and strains all vanish.

Why have we done all this? To show how a positive definite stored energy function places constraints on the possible values of the elastic constants. The first step is to expand the stress and strain matrices in (2.26) into deviatoric and volumetric parts. Using the definitions (2.22) and (2.23), we can rewrite (2.26) as

$$W = \frac{1}{2} \, tr\left\{\left(\boldsymbol{\sigma}^d + \frac{1}{3} I_1 \mathbf{1}\right)\left(\boldsymbol{\epsilon}^d + \frac{1}{3} e\mathbf{1}\right)\right\}$$

$$= \frac{1}{2} \, tr\left\{\boldsymbol{\sigma}^d \boldsymbol{\epsilon}^d + \frac{1}{3} I_1 \boldsymbol{\epsilon}^d + \frac{1}{3} e\boldsymbol{\sigma}^d + \frac{1}{9} I_1 e\mathbf{1}\right\}.$$

The trace operator can be taken inside the brackets and applied to each term individually. If we do this we find

$$W = \frac{1}{2} \, tr(\boldsymbol{\sigma}^d \boldsymbol{\epsilon}^d) + \frac{1}{2} \left(\frac{1}{3} I_1 e\right) \tag{2.27}$$

where we used the fact that $tr\boldsymbol{\sigma}^d = tr\boldsymbol{\epsilon}^d = 0$. Note that even though both $\boldsymbol{\sigma}^d$ and $\boldsymbol{\epsilon}^d$ are traceless, the product $\boldsymbol{\sigma}^d \boldsymbol{\epsilon}^d$ may well have a nonzero trace. Equation (2.27) shows that W consists of two parts: the stored energy associated with volumetric deformation given by one-half the product of the mean stress with the volumetric strain and the stored energy associated with distortion given by one-half the trace of the product of stress and strain deviators.

Next, we can replace $\boldsymbol{\sigma}^d$ and $(1/3)I_1$ in eq. (2.27) using (2.21) and (2.24). That gives

$$W = G tr(\boldsymbol{\epsilon}^d)^2 + \frac{1}{2}\left(\lambda + \frac{2}{3} G\right)e^2. \tag{2.28}$$

Here both $tr(\boldsymbol{\epsilon}^d)^2$ and e^2 will be positive, or may be zero, but only if the strains themselves are zero. Also, W is, by agreement, positive definite. It can be zero only if all the strains vanish. Also $\boldsymbol{\epsilon}^d$ and e are independent. We can arbitrarily alter one without affecting the other. These things taken together imply that

$$G > 0 \quad \text{and} \quad \left(\lambda + \frac{2}{3} G\right) > 0 \tag{2.29}$$

giving our first constraints on the constants.

Now we can use the relationships found in Table 2.1 to see the implications of these two inequalities. First, since K is equal to $\lambda + (^2/_3)G$ we immediately have

$$K > 0. \tag{2.30}$$

We can use the relationship

$$E = \frac{G(3\lambda + 2G)}{\lambda + G}$$

to investigate Young's modulus. Clearly the numerator of this fraction is strictly positive. The denominator is also positive since

$$\lambda + G = \lambda + \frac{2}{3}G + \frac{1}{3}G > 0.$$

Thus, both numerator and denominator are positive and

$$E > 0. \tag{2.31}$$

Finally, we must investigate Poisson's ratio. We have this relation for ν

$$\nu = \frac{\lambda}{2(\lambda + G)}. \tag{2.32}$$

From eq. (2.29) we see that $\lambda > - (^2/_3)G$. Using this in (2.32) shows ν must be greater than -1. The greatest value λ can take on is infinity. It takes this value for materials that are incompressible. If we take the limit of the right-hand side of (2.32) as $\lambda \to \infty$, we find ν equals $^1/_2$. Thus, the bounds on Poisson's ratio are

$$-1 < \nu \le \frac{1}{2}. \tag{2.33}$$

Together, eqs. (2.29), (2.30), (2.31), and (2.33) constrain the values of the elastic constants. These constraints will be useful in establishing the particular value of a constant for a given material, the subject of the remainder of this chapter.

One final comment to end this section concerns the lower bound on Poisson's ratio found in eq. (2.33). If we reflect for a moment on the definition of

Poisson's ratio from eq. (2.2), we see that negative values of ν seem to be physically unlikely. If we were in possession of an elastic material whose Poisson's ratio was negative and we were to stretch a bar of this material, then it would grow fatter rather than thinner. In fact, materials do exist with negative Poisson's ratio, but they are extremely rare and would never be encountered in day-to-day engineering practice. For this reason we can safely assume a lower bound of zero for ν for engineering purposes.

2.8 Determination of elastic constants

To determine values for the elastic constants for a given soil deposit, we will need to perform a test or perhaps several tests. We can broadly categorize all tests as either *laboratory tests* or *field tests*. Both have certain advantages and certain disadvantages.

The great advantage of laboratory tests is that we are able to control the conditions of the test quite carefully. We can arrange to test the soil in different configurations and can control the stresses and deformations with great accuracy. This is not often possible to accomplish in field tests. The disadvantage of laboratory testing is that all tests will be performed on samples that have been removed from the actual soil deposit in the field. Thus, it is highly likely the soil we are testing in the laboratory will be different in some (perhaps important) respects from the soil in the field. Sample disturbance and its consequences is a field that is not well-understood. Experience has shown that many soils are extremely sensitive to sampling disturbance and laboratory test results may be almost worthless. Other soils may perform quite well during sampling and yield reliable laboratory data.

The most important factor regarding sampling disturbance is the quality of sampling technique. Samples may be taken from a bore hole by a wide variety of methods. For fine grained soils, devices range from split barrel samplers (which cause a great deal of sample disturbance) to the Swedish foil continuous sampler (which may induce very little disturbance). Coarse grained soils offer even more severe sampling problems, first because the grain size may be too big to permit sampling by conventional methods, second because the absence of cohesion will cause sample disintegration. Ground freezing methods can be used to overcome some problems with coarse grained soils, but this can be a very expensive procedure. Even when a good sample is removed from a borehole, disturbance may still occur during transportation to the laboratory, in removal of the sample from the sampling device, or during preparation of the sample for testing.

Naturally, none of these problems arise in field testing. Field tests involve in-situ loading of the soil and, especially for sensitive soils, this is a great advantage. Field-test methods range from relatively simple penetration sounding to complex and expensive plate load tests. Expense is one of the disadvantages associated with field tests. Less-expensive field tests will yield results with significant amounts of uncertainty. To reduce the degree of uncertainty, it is necessary to use more sophisticated tests, which rapidly increase in expense.

Field-test methods suffer other disadvantages in addition to expense. All field tests involve loading a certain volume of the in-situ soil, and in most circumstances this volume is not precisely known. Whether or not natural inhomogeneities may affect test results cannot be known with much certainty. Also, pore pressures may be generated by the test load, and these may significantly distort the results. This is particularly true in fine grained soils but may also occur in sands when relatively rapid loading, such as penetration testing, is used. A third disadvantage of field testing is that no single test will deliver good-quality values for two elastic constants. Some tests provide a single constant; for example, the pressuremeter test, which gives the shear modulus. Other tests provide a value for an algebraic combination of two constants, but more information is needed to find their separate values. Plate-loading tests fall into this category, as do certain geophysical test methods.

At the end of the day, there are no simple answers to the question: How should we determine two elastic constants for a given soil deposit? You, the geotechnical engineer, will decide on appropriate test methods based on knowledge of what test equipment may be available, how much money has been budgeted for testing, the applications that test results will be used for, and whatever is already known concerning the soils involved.

2.9 Laboratory tests

When we wish to determine *appropriate* values for elastic constants using laboratory tests, it will always be necessary to make a precise definition of what we mean by the word appropriate. As we know, soil is neither linear nor elastic, and we use elastic theory for reasons solely to do with utility – because it works. To make it work, we need to plug the right numbers into the spots where the elastic constants occur in the equations. For all the common laboratory tests, nonlinear stress versus strain response will occur, and we will be forced to choose a point or points from the test data at which the calculation of an elastic constant will be made. Usually we will be interested in the slope of a line on a stress-strain diagram, and this slope will give us one or another of the elastic moduli. The big question is this. Where and how should I define this slope? Should it be tangent to the curve or should it be a secant? Should it be near the beginning of the curve, where the test strains are still very small, or should it be later in the test after the stress has increased to a significant value? These are very important questions, and there are no hard and fast answers. Like many questions in geotechnical engineering, the only general thing we can say is that we must do things in an *appropriate* way. We must find elastic constants which, for the problem we are faced with, will give us roughly correct answers. So the word "appropriate" involves not only the soil and the test but also the application we have in mind. The selection of elastic constants for a particular soil and a particular engineering problem will involve all the engineer's skills. Experience and judgment will play as important a role as does knowledge of elasticity theory.

A relatively long history of soil testing exists, with the earliest definitive testing data from the late nineteenth century. In 1883, George Darwin, a British engineer (and nephew of Charles Darwin), presented test results on sand that indicated how the arrangement of grains (and hence the relative density) affected strength. Until that time, engineers had used the angle of repose to characterize sand strength. Not long after, an Irish engineer, A. L. Bell, performed rudimentary shear strength tests on clay. These two investigations were the beginning of a long tradition in soil testing that concentrates attention on strength. It was not until relatively recently that interest has grown in determination of elastic response of soil. Sometimes traditional methods developed for strength determination will also give information on elastic constants. This is true for triaxial testing. Other cases, such as the direct shear test, give little insight into elastic behavior.

The triaxial test employs a cylindrical sample of soil encased in a rubber membrane. The sample is placed in a pressure cell and a hydrostatic stress is applied. Sample volume change due to the hydrostatic stress can be detected by various methods. The easiest method involves using a fully saturated sample and simply measuring how much pore fluid is expelled from the sample for a given change in cell pressure. Typical results are illustrated in Figure 2.8. The cell pressure is denoted by σ_3 (since it will become the minor principal stress shortly), and, since this test is characterized by compression, we make compressive stress and strain positive. In general, the cell pressure-volumetric strain response will be nonlinear. The slope of the experimental curve at the appropriate stress level can be used to estimate the bulk modulus K. The bulk modulus will generally be sensitive to sample disturbance. For samples of fine-grained, fully saturated soils, time will be required for consolidation to occur under each increment of cell pressure.

When any soil is subjected to compressive stress, it may undergo irreversible

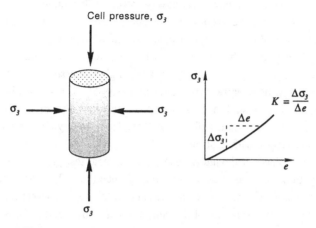

Figure 2.8 Hydrostatic compression phase of triaxial test.

volumetric strains. These occur because of rearrangement of the grain structure representing inelastic behavior. A loose, freshly deposited soil will suffer these irreversible strains each time the stress is increased. This might occur because of accumulation of overburden as new soil is deposited on top of old. Each stress increase causes some grain rearrangement and, in effect, a new soil particle structure. If subsequently the stress is reduced (as might happen if overburden is removed by erosion), this rearranged structure is locked into the soil, and the soil will behave differently if reloaded. These concepts are described by the terms *normally consolidated* and *over-consolidated*. Normally consolidated soils are those deposits in which the present level of stress is the greatest that has existed throughout the soil's history. No unloading due to erosion or other causes has occurred, and the particle structure corresponds to the present stress level. In contrast, over-consolidated soils have been subjected to higher stress levels at some time in the past, but unloading has occurred and the particle structure is more dense than would be expected based only on knowledge of the present day in-situ stresses. A common cause for over-consolidation is glaciation. Soils in many parts of the world have been subjected to quite large stresses due to ice loading, which has subsequently disappeared as glaciers retreated.

The distinction between normally and over-consolidated states is important in relation to bulk modulus determination. An over-consolidated soil subjected to high stress levels may exhibit relatively elastic response in hydrostatic compression. The response may not be linear, but it will be reversible or nearly so. Many over-consolidated soils will be relatively insensitive to sample disturbance, and good estimates for bulk modulus may be found. In contrast, normally consolidated soils may not behave so nicely. When an undisturbed sample is removed from the ground, the stress is immediately reduced to near zero. If it is then subjected to hydrostatic compression, we can anticipate finding more or less reversible behavior up to the level of stress that existed in the ground at the point of sampling; but for higher stresses, irreversible deformation and particle rearrangement must be expected. Unfortunately, we will usually be most interested in the soil response at loads near or slightly above the in-situ stress level. Values of K determined for normally consolidated soils must be approached with caution.

The second phase or deviatoric phase of a triaxial test provides more useful information concerning elastic constants. Once the cell pressure has been applied, an axial force is placed on the sample by way of a vertical spindle that protrudes through the top of the cell. The axial stress in the sample is increased while the radial stress is held constant at the cell pressure, as illustrated in Figure 2.9. The increase in axial stress is denoted q and called the *deviatoric stress*. The axial stress becomes the major principal stress σ_1, and $\sigma_1 = \sigma_3 + q$. Young's modulus is the slope of the q versus axial strain curve. We will denote the axial strain by ϵ_1. Since the $q - \epsilon_1$ curve may be nonlinear even for small values of q, conventional practice takes E to be the slope of the secant that joins the origin to a particular point on the curve. This point corresponds to the working stress level and is often taken to be one-third the maximum value for q found later in the test. We can expect the value determined for Young's modulus to be sensitive to sample disturbance.

At this point, we need to distinguish between drained and undrained test procedures. Drained tests are tests in which the pore fluid is allowed to freely move into and out of the sample. In the case of clays, this may involve long periods of time to accomplish, and careful drained tests may take several days to perform. Undrained tests are those in which no pore fluid is allowed to enter or leave the sample, either because drainage is physically blocked, or because time is not allowed for drainage to occur. In an undrained test, the value of E that we determine will be denoted E_u. It may be quite different from the value found in a drained test. We will discuss the significance of E_u, and of all undrained deformations, at the end of this chapter. Undrained deformations comprise a particular type of response called *incompressible behavior*. They play an important role in our understanding of soil behavior. If the test is a drained test, on the other hand, we will denote Young's modulus simply by E. In general, E and E_u will be different, as will be discussed in Section 2.11.

There is one more piece of information that may be obtained from a drained triaxial test: Poisson's ratio. We can find Poisson's ratio, provided the sample volume change is measured during application of the deviatoric stress. Vol-

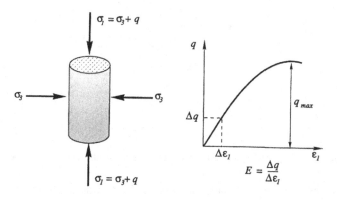

Figure 2.9 Deviatoric phase of triaxial test.

ume change measurement allows us to calculate the volumetric strain e. We can plot e versus ϵ_1, the axial strain, to find a curve that typically has the shape shown in Figure 2.10. Compressive strain is considered positive here. If we denote the radial strain in the cylindrical sample by ϵ_3, then the volumetric strain is given by

$$e = \epsilon_1 + 2\epsilon_3$$

which follows from eq. (1.10). Poisson's ratio is defined by

$$\nu = -\epsilon_3/\epsilon_1$$

and combining these two equations we find

$$\nu = \frac{1}{2}(1 - e/\epsilon_1). \tag{2.34}$$

In all triaxial tests, the e versus ϵ_1 response will be nonlinear. Indeed, for many soils, dilation effects will make e negative (indicating an increase in volume) when the deviatoric stress nears its maximum value. Even in the early stages of a test, significant nonlinearity may occur. Conventional practice suggests we define Poisson's ratio at the point on the $e - \epsilon_1$ curve that corresponds to the point on the $q - \epsilon_1$ curve where E was determined. Comparing Figures 2.9 and 2.10, the value of $\Delta\epsilon_1$ used in Figure 2.9 will be the same as in Figure 2.10. Poisson's ratio is determined by evaluating eq. (2.34) at this point, as shown on the figure.

The elastic constants

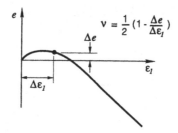

Figure 2.10 Typical volumetric strain versus axial strain response in triaxial test.

The value of Poisson's ratio, and especially the value of Young's modulus, determined from triaxial tests will be dependent on the cell pressure used in the test and on the degree of over-consolidation that exists in the sample. In general, Young's modulus will increase with increasing cell pressure. Poisson's ratio may increase or decrease or remain relatively constant depending upon the particular soil involved, but in most cases, a small decrease with increasing cell pressure is observed. Figure 2.11 shows data from drained triaxial tests on a uniform beach sand that exhibit these effects.

A second common laboratory test that can offer information concerning elastic constants is the oedometer test (sometimes called the consolidation test). In this test a thin cylindrical sample is subjected to *uniaxial compression* as illustrated in Figure 2.12. A vertical load is placed on the sample, and it is constrained from expanding horizontally by a rigid ring. All deformation is vertical and the only nonzero strain is the vertical compressive strain denoted ϵ_1. The strain and stress matrices will look like this:

$$\boldsymbol{\sigma} = \begin{bmatrix} \sigma_1 & 0 & 0 \\ 0 & \sigma_3 & 0 \\ 0 & 0 & \sigma_3 \end{bmatrix}, \quad \boldsymbol{\epsilon} = \begin{bmatrix} \epsilon_1 & 0 & 0 \\ 0 & 0 & 0 \\ 0 & 0 & 0 \end{bmatrix} \quad (2.35)$$

where σ_1 is the vertical stress due to the load P. We see that the volumetric strain e is equal to the axial strain ϵ_1, so that Hooke's law in the form (2.19) tells us

$$\sigma_1 = \lambda e + 2G\epsilon_1 = (\lambda + 2G)\epsilon_1.$$

Using the relationships between the elastic constants we find

$$\frac{\sigma_1}{\epsilon_1} = \lambda + 2G = \frac{E(1 - \nu)}{(1 + \nu)(1 - 2\nu)}. \quad (2.36)$$

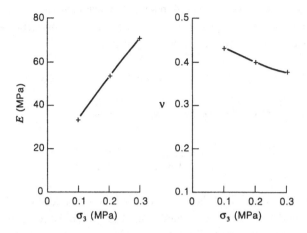

Figure 2.11 Actual test data for dependence of E and ν on cell pressure for the sand described in Figure 1.1.

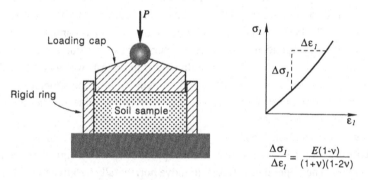

Figure 2.12 Oedometer test configuration.

In general, the σ_1 versus ϵ_1 response will be nonlinear, and eq. (2.36) must be applied to some appropriate part of the curve, as shown in Figure 2.12. Note that a single elastic constant is not determined here, but instead we find an algebraic combination of two constants. This is an unfortunate feature of many tests, in that other tests are necessary to provide more information before individual constants can be determined. Looking at eq. (2.36) one might be tempted to think a good estimate for E can be found by making a guess for the value of ν. After all, we know ν must lie between zero and one-half, so perhaps a guessed value isn't such a bad idea. Unfortunately, this idea does not work well. The value of E determined from (2.36) is quite sensitive to the value chosen for ν. For example changing ν from $1/4$ to $1/3$ changes E by 30%.

We can make one other observation about the oedometer test. The rigid ring surrounding the sample applies the horizontal stress component σ_3, which, from Hooke's law and eq. (2.35), will be given by

$$\sigma_3 = \lambda \epsilon_1.$$

We see that the ratio of horizontal to vertical stress is

$$\frac{\sigma_3}{\sigma_1} = \frac{\lambda}{\lambda + 2G} = \frac{\nu}{1 - \nu} \tag{2.37}$$

where the second equality follows from the relations between the elastic constants. We note here that if one additional measurement were made, σ_3, then we could obtain estimates for both E and ν from a single oedometer test. Unfortunately, the radial stress is not easily measured, and the additional information is rarely obtained. This illustrates a universal fact: A single set of stress and strain data can, at best, determine only one elastic constant. It may determine an algebraic combination of two constants, but it cannot determine two constants independently. In order to determine two constants we need two independent sets of data. Thus, in a drained triaxial test, we obtain E and ν by having two data sets: q versus ϵ_1 and e versus ϵ_1. In an oedometer test, we have only one data set, σ_1 versus ϵ_1, and we obtain a value for the quantity $E(1 - \nu)/(1 + \nu)(1 - 2\nu)$.

The other common laboratory strength test is the direct shear test, illustrated in Figure 2.13. In this test, the sample is placed in a rigid box having two halves. The upper half is forced to move horizontally, relative to the lower half, and shear stress is generated on the soil surface in the plane between the two halves of the box. The deformation is far from homogeneous, and discontinuities, accompanied by stress concentrations, occur in the soil near the walls of the box. Rough estimates for the shear modulus G can be obtained as shown in Figure 2.13 but, because of the nonuniformity of both stress and strain fields, these results must be used with caution.

There is a growing number of other laboratory test configurations that can be used to supply values for elastic constants, but these are all mostly used in research and will rarely be available in ordinary practice. Pure shear tests, true triaxial tests, and resonant column tests may be found in many research laboratories, but their use is not yet sufficiently widespread to be encountered by most engineers. The triaxial test remains our best tool for estimating elastic constants in the laboratory.

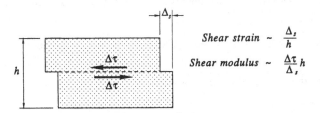

Figure 2.13 Direct shear test geometry.

2.10 Field tests

When values for elastic constants are wanted, field tests are the second string to the geotechnical engineer's bow. Field testing can be used as a complement to laboratory tests or can be used exclusively when sampling and sample disturbance appear to present significant difficulties for laboratory testing. We can identify two categories of field tests, *correlative tests* and *deterministic tests*.

The correlative test category uses simple test procedures whose results, by virtue of experience, have been correlated to various soil properties including elastic constants. The most common correlative tests are penetration soundings. The standard penetration test (SPT) and cone penetration test (CPT) are familiar examples. Both of these tests involve forced penetration of the ground by either a standardized impact loading (SPT) or by a static force (CPT). The SPT test employs a split barrel sampler with standard dimensions, and the number of hammer blows required to produce a certain amount of penetration is counted. Test results are sensitive to details of the test rig, as well as the soil being tested and some degree of uncertainty must accompany any SPT test. The CPT test appears to produce more consistent results. In it, a steel cone of standard dimensions is forced into the ground by a hydraulically controlled static load. A uniform rate of penetration is used, and the force needed to cause penetration (called the cone resistance) is measured continuously.

Experience has shown that results from both SPT and CPT testing can be vaguely correlated to elastic constants. Most researchers who have looked for these correlations agree on one thing: that a linear correlation is appropriate. That is, the SPT or CPT test result is thought to be a linear function of elastic properties of the soil tested. This linearity is probably as much an indication of the uncertainty involved in the correlation as it is a reflection of the physics involved. If one has little definitive data, then the simplest possible form (in this case linear) is probably best. Two widely used correlations for sandy soils are

(i) Parry's (1978) SPT correlation: If we use N to define the number of blows necessary to produce a penetration of one foot (or 300 mm in SI notation) in an SPT test, then Parry suggested

$$\frac{N}{3} = \frac{E}{1 - v^2} = \frac{2G}{1 - v}$$

where E has dimensions of MPa. The reason we find a combination of elastic constants on the right-hand side results from Parry's interest in a particular foundation problem for which this combination was intrinsically valuable. We'll find the ratio $2G/(1-v)$ cropping up in Chapter 4 when we consider settlement of foundations. Also, it should be noted here that the value of N used is the so-called corrected standard penetration. Empirical corrections are employed to account for different overburden stresses found at different depths encountered in SPT testing.

(ii) Schmertmann's (1970) CPT correlation: The cone penetration resistance is usually abbreviated q_c and is usually reported in units of kg/cm^2. Schmertmann suggested the following correlation

$$\frac{q_c}{5} = \frac{(1 - v)E}{(1 + v)(1 - 2v)}$$

where q_c is in kg/cm^2 and E in MPa. Again we find a combination of elastic parameters, this being the form discovered for uniaxial compression in eq. (2.36).

On page 68 we will discuss how elastic constants can be determined by measuring the velocity of elastic waves. One particular type of wave, the transverse or S-wave, depends on the shear modulus, and its velocity, denoted c_T, is given by $\sqrt{G/\rho}$. If c_T and the soil density ρ are known, then G follows from

$$G = \rho c_T^2.$$

Earthquake engineers are often interested in the S-wave velocity in shallow soil deposits because it plays an important role in how seismic waves radiating upward from basement rock may be modified before reaching the ground surface. Japanese engineers have particularly been interested in finding ways to estimate c_T, and they have studied correlations between c_T and other variables. The most common correlation gives c_T in terms of the standard penetration value N, measured in blows per foot. Usually the raw or uncorrected N value is used. Most Japanese studies have found relationships of this form

$$c_T = aN^b$$

where a and b are constants. We immediately see the implication

$$G = \rho a^2 N^{2b}.$$

Typically a value for b is found near 0.35, suggesting that G and N may not be linearly correlated after all. The most complete study of c_T versus N correlations is by Ohta and Goto (1978). In their study, they correlate c_T not only to N but also to depth below the ground surface, soil type (clay, fine sand, medium sand, coarse sand, sandy gravel, and gravel), and geologic age (holocene or pleistocene). By including more variates, they find the correlation can be markedly improved.

In contrast to correlative testing, deterministic tests will generally give estimates for elastic constants with greater reliability, although generally at greater expense. Most of the deterministic test methods involve static loading, but dynamic methods can also be effective. The most direct type of deterministic field test is the *plate load test*. In this test, a rigid steel plate is placed on the soil surface and subjected to a vertical load, as illustrated in Figure 2.14. The plate may be circular or square and will generally have overall dimensions on the order of 300 mm. Load is applied by a hydraulic jack using a grouted anchor or anchors as reaction to jack against. As the total load $\mathcal{P}$ is increased, the vertical displacement δ of the plate is measured. We can use the measured values of $\mathcal{P}$ and δ to directly infer the elastic properties of the soil provided we assume the soil responds elastically. The relationship for a circular plate is

$$\frac{\mathcal{P}}{2a\delta} = \frac{E}{1 - \nu^2} = \frac{2G}{1 - \nu} \tag{2.38}$$

where a is the plate radius. Note that this is not a correlation; it is an exact result from the theory of elasticity, and we will find it again in Chapter 4. Note also that we obtain an algebraic combination of two constants rather than a single constant. Finally, and perhaps most importantly, note that if we plot $\mathcal{P}$ versus δ for a typical test, we will not get a straight line. The old nonlinearity problem raises its head again with deterministic field tests, just as it did

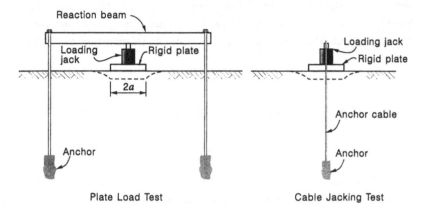

Plate Load Test   Cable Jacking Test

Figure 2.14 Plate load test geometries.

in the laboratory. And as with lab tests, we will need to select an *appropriate* load level $\mathscr{P}$ at which to make the calculation in eq. (2.38) and that appropriate level will depend on the application as well as the test method and the soil involved.

A secondary assumption used in obtaining eq. (2.38) is that the base of the plate is smooth, allowing horizontal displacements of the soil in contact with the plate. Occasionally, plate bearing tests are performed with a cast-in-place concrete pad rather than a smooth steel plate. In these cases, the assumption of smooth contact isn't very good. For a rough contact (implying zero horizontal displacement beneath the plate) eq. (2.38) is replaced by

$$\frac{\mathscr{P}}{2a\delta} = \frac{E \, \ell n(3 - 4\nu)}{(1 + \nu)(1 - 2\nu)}. \tag{2.39}$$

This is also an exact solution to the elasticity problem. If the plate is square rather than circular in shape, then no exact solution can be found. An approximate solution is available, however. It gives

$$\frac{\mathscr{P}}{2a\delta} = \frac{\chi E}{1 - \nu^2} \tag{2.40}$$

where χ is a constant, equal to 0.877, and the plate width is $2a$.

A similar method to the plate load test is the *screw plate test*. This test is illustrated in Figure 2.15. A helical-shaped plate is mounted on a steel shaft that can be augered into the ground (provided the soil is not too hard). Once

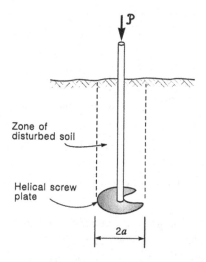

Zone of
disturbed soil

Helical screw
plate

2a

Figure 2.15 Screw plate test method.

in place, a vertical load $\mathcal{P}$ is applied and the displacement δ measured. The loading geometry can be idealized as a rigid circular plate buried in the soil and subjected to a vertical load. The problem was studied by Selvadurai and Nicholas (1979) who found that the undrained Young's modulus could be approximately related to $\mathcal{P}$, δ, and the plate radius a as

$$E_u = \frac{\zeta\mathcal{P}}{\pi a\delta} \tag{2.41}$$

where ζ is a constant whose value is near 0.60. The actual value of ζ depends (unfortunately) on Poisson's ratio and on the details of the idealized model one uses to represent the screw plate. We will examine one particularly simple model for this problem in Chapter 4.

The *borehole plate load test* is very similar in concept to the screw plate test. A circular plate is subjected to vertical load at the base of an uncased borehole as shown in Figure 2.16. A value for Young's modulus can be obtained, having the same form as eq. (2.41), but with ζ taking a value near 0.50 rather than 0.60. The exact value of ζ depends as above on the value of Poisson's ratio and on the modelling details.

A second test that can be used when an uncased borehole is available is the *pressuremeter test*. A device that is basically a calibrated cylindrical balloon is placed down the bore and then inflated. It exerts a normal traction on the borehole walls as shown in Figure 2.17. A value for the shear modulus G can

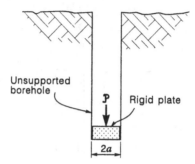

Figure 2.16 Bore hole plate test.

be obtained from this test, provided we make some assumptions. We need to assume that drilling has not introduced an appreciable amount of disturbance in the soil surrounding the bore and that the displacements caused by inflating the balloon are horizontal. This second assumption is valid in the region near the center of the balloon, and because of this pressuremeter measurements are designed to reflect the soil response near the center of the loaded region. The shear modulus G is directly related to the inflation pressure p and the radial displacement u_r of the borehole wall as shown in Figure 2.17. In older versions of pressuremeter apparatus the displacement u_r was not measured, but the volume of the balloon was monitored, and, if the initial volume was known, volume could be used instead of u_r to determine G. This is also shown on Figure 2.17. The pressuremeter is a useful test that generally gives reliable results. It is also an interesting loading geometry, and we will study it again in Chapter 4.

The final category of deterministic field tests we will consider is *wave propagation testing*. Small amplitude waves can be generated in the ground and their velocities measured. If the ground behaves elastically, the wave velocities will depend on the elastic constants and the soil density ρ, and values for the constants may be directly inferred. There are question marks here, however, concerning interpretation of wave propagation results, just as with all of the other tests we have discussed. In most instances, wave propagation methods will generate only very small strains in the soil, far smaller than may be *appropriate* to the engineering problem we wish to consider. As a result, the values of elastic moduli inferred from wave propagation tests may be significantly larger than values obtained by other test methods such as plate loading tests. And so, once again we find ourselves in a position where engineering judgment is extremely important.

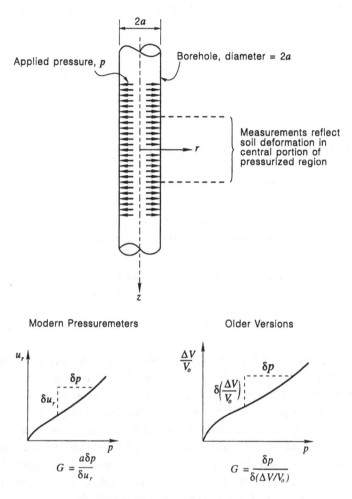

Figure 2.17 Geometry of pressuremeter test.

There are two broad classes of elastic waves called *body waves* and *surface waves*. Body waves are those capable of propagating through the interior of an elastic body, while surface waves require the presence of a surface, either a free surface or an interface between two different materials. The theory of elasticity shows there are exactly two types of body waves: longitudinal waves or P-waves (primary waves), and transverse waves or S-waves (secondary waves). There are several types of surface waves, but only one has relevance to soil testing. It is the Rayleigh wave. Each type of wave is char-

The elastic constants

acterized by a different velocity. The propagation velocities for body waves
are

$$\text{P-wave velocity:} \quad c_L = \sqrt{\frac{E(1 - \nu)}{(1 + \nu)(1 - 2\nu)\rho}}$$

$$\text{S-wave velocity:} \quad c_T = \sqrt{\frac{G}{\rho}}$$

(2.42)

while the Rayleigh wave velocity is given by

$$\text{Rayleigh wave velocity:} \quad c_R = \alpha c_T \qquad (2.43)$$

where α is a constant that depends on the value of Poisson's ratio. It can be
found as a root of the following polynomial equation:

$$\alpha^6 - 8\alpha^4 + 8\left(3 - \frac{1 - 2\nu}{1 - \nu}\right)\alpha^2 - 16\left(1 - \frac{1 - 2\nu}{2 - 2\nu}\right) = 0. \qquad (2.44)$$

It can be shown that only one root of this equation will be admissible (real
and positive).

At first glance, the complexity of eq. (2.44) would suggest that the Rayleigh
wave velocity might not be particularly useful for determining elastic constants, but in fact the value of α is constrained to lie in a narrow range. For
Poisson's ratio equal to 0.25, the value of α is 0.919. For greater Poisson's
ratio, α increases, equalling 0.955 for $\nu = 0.50$ (the largest possible value).
So we see α is close to 1.0 and c_R is close to c_T. For many applications, c_R
and c_T will be assumed to be equal. Then G follows from the second of (2.42),
provided a value for the density ρ is known. Generally a good estimate for ρ
will be available. If the P-wave velocity is known, only the combination of E
and ν found in the first of (2.42) [as well as in (2.36)] can be determined. Of
course, if both c_L and c_T are known, then we have sufficient information to
find all the elastic constants.

There are various ways to go about generating these waves and finding their
velocities. We can identify two general categories of tests, those performed
on the ground surface and those performed underground using a borehole.

The most simple test at the ground surface is to use an impulse (such as hit-
ting the ground with a hammer) to generate waves, and then placing two sensors on a radial line from the point of impulse. The waves propagate outward
and the time delay between arrivals at the sensors is accurately measured. The
wave velocity is found by dividing the distance between sensors by the time

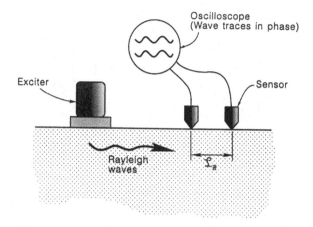

Figure 2.18 Surface wave generator to measure Rayleigh wave length.

delay. The impulse will generate a P-wave and possibly S-wave and Rayleigh waves as well, although these may not be so distinctly shown on the record of ground motion. A second method that is not so simple, but generally gives more reliable results, is to use steady-state excitation rather than an impulse. An oscillator, usually electrically driven, is located on the ground surface, and two sensors are placed on a radial line, as in the impulse method. The oscillator will generate a continuous train of Rayleigh waves in the soil. With the steady-state method, we can move the sensors while the test is in progress. Holding one sensor stationary, the second sensor is moved until the wave motion sensed by both sensors comes into phase, as illustrated in Figure 2.18. The separation between the sensors is then the wave length of the motion, $\mathscr{L}_R$. The Rayleigh wave velocity is found by multiplying $\mathscr{L}_R$ by the excitation frequency.

In the second category of tests, one or more boreholes are used. Sensors can be placed at the ground surface or down the bore, and an impulse is used to generate P- and S-waves. The impulse can be applied at the bottom of the bore and the sensor placed on the ground surface. The time delay between the impulse and the arrival of the wave at the sensor is measured. This is called an up-hole test. The impulse can be generated by an SPT test. The reverse configuration, with the sensor placed in the bore and the impulse applied at the surface, is called a down-hole test. A third type of test is the cross-hole test in which two or more bores are used. Both the impulse and the sensor are located in separate bores, and the wave path is horizontal rather than vertical. All three types of test are summarized in Figure 2.19. In each case, the soil tested is that lying on the propagation path.

The elastic constants

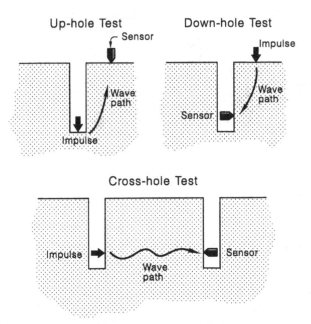

Figure 2.19 Borehole wave velocity tests.

Of all the field testing methods, no particular one can be singled out as prefer-able to the others. Each test has its advantages, but no test is without disadvan-tages, and the selection of which test or tests to use will usually be a matter of availability and economics. Generally speaking, the larger the testing budget is, the more reliable will be our estimates for the elastic constants. It is an unfortu-nate fact in geotechnical engineering that clients are, more often than not, un-willing to invest much in site investigations. The general attitude of the client is this: "I don't want to spend money on something I can't see." Boreholes, plate-load tests, and triaxial tests are invisible to most clients. Occasionally the fortu-nate geotechnical engineer may encounter an enlightened client who realizes care-ful investigations early in a project may save large sums later, but this is the exception rather than the rule. This does not decrease the importance of careful analysis and design, but we must be aware that in most situations there will be significant uncertainty associated with our knowledge of the elastic constants.

2.11 Incompressible elasticity

To end this chapter, we will point out an important special category of elas-ticity problems that have relevance in geotechnical engineering. These are problems involving incompressible elastic bodies. Incompressibility, we realize, is an idealized property. Any material, no matter how hard or strong, will compress if we subject it to increasing pressure. But for certain materials, many types of rubber for example, the bulk modulus K is much, much greater than the shear modulus G, and shearing deformations will be much more important than dilatation or compression in the overall response of the body. The behavior of these bodies can be closely approximated by assuming them to be incompressible.

Incompressibility is relevant to soil mechanics whenever we wish to consider the response of a fully saturated soil in *undrained conditions*. By undrained we mean that for some reason or other, the pore fluid cannot freely move inside the matrix of solid particles. This might be because we have turned off the drainage tap in a triaxial test, or it might be that there has simply not been time enough for pore fluid movement to occur. Consolidation problems generally begin as undrained deformations, but change to drained deformations as time permits pore water to escape.

We realize that both the solid particles and the pore water are slightly compressible, but in a geotechnical context it is perfectly sensible to assume they are incompressible. Volume changes that occur in soil are the result of rearrangement of particles. Only a tiny fraction of the total volume change exhibited by a soil results from compression or expansion of individual particles. Compression (or extension) of a soil reflects an overall rearrangement of the positions of individual particles and an accompanying decrease (or increase) in the pore volume. If the pores are fully occupied by water, then flow must occur in order for pore volume to change. So in undrained conditions, when no flow occurs, the soil is virtually incompressible.

An incompressible body is characterized by infinite bulk modulus. If we examine Table 2.1, we see that $K \to \infty$ implies $\lambda \to \infty$ and $\nu \to 0.5$. Also $E \to 3G$. For an incompressible body we only need to specify either E or G. The other three constants are known. Sometimes books and articles on geotechnical subjects will refer to the so-called *undrained Young's modulus* E_u. If we realize that the shear modulus simply reflects the distortional stiffness of the soil and isn't affected by the presence or absence of pore fluid (which has no shear strength), then we see that G is the same in drained or undrained conditions and $E_u = 3G$. Moreover, if we know the values of any two elastic constants for a soil in drained conditions, then we can find G and hence E_u. In Chapter 4 we will consider some undrained deformations that will be characterized by incompressible behavior.

Exercises

2.1   Consider the elastic bar subjected to uniaxial tension (as shown in Figure 2.1). Given that

$$\sigma = \begin{bmatrix} \sigma_{xx} & 0 & 0 \\ 0 & 0 & 0 \\ 0 & 0 & 0 \end{bmatrix}$$

and that eq. (2.19) applies, but with no other information, prove the following

(i) $\epsilon_{xy} = \epsilon_{yx} = \epsilon_{yz} = \epsilon_{zy} = \epsilon_{zx} = \epsilon_{xz} = 0$

(ii) $\epsilon_{yy} = \epsilon_{zz}$

(iii) $\dfrac{\epsilon_{yy}}{\epsilon_{xx}} = - \dfrac{\lambda}{2(\lambda + G)}$

(iv) $\dfrac{\sigma_{xx}}{\epsilon_{xx}} = \dfrac{G(3\lambda + 2G)}{\lambda + G}.$

2.2   Given eqs (2.11), (2.15), and (2.18), prove the following

(i) $\lambda = \dfrac{3K\nu}{1 + \nu}$

(ii) $K = \dfrac{GE}{3(3G - E)}$

(iii) $G = \dfrac{3KE}{9K - E}$

(iv) $\nu = \dfrac{\lambda}{2(\lambda + G)}$

(v) $E = \dfrac{G(3\lambda + 2G)}{\lambda + G}.$

2.3   (i) Derive an expression for the stored energy function W as a function of E, ν, and applied stress σ_1 for the stress and strain conditions in eqs. (2.35) found in an oedometer test.
      (ii) Derive an expression for W as a function of E, ν, σ_1, and σ_3 for a conventional triaxial test.

2.4 For the conventional triaxial compression test, the stress matrix is given by

$$\boldsymbol{\sigma} = \begin{bmatrix} \sigma_1 & 0 & 0 \\ 0 & \sigma_3 & 0 \\ 0 & 0 & \sigma_3 \end{bmatrix}.$$

Show that the corresponding *deviator* stress matrix is

$$\boldsymbol{\sigma}^d = (\sigma_1 - \sigma_3)\begin{bmatrix} 2/3 & 0 & 0 \\ 0 & -1/3 & 0 \\ 0 & 0 & -1/3 \end{bmatrix}$$

which suggests why we call $q = \sigma_1 - \sigma_3$ the *deviatoric* stress.

2.5 A conventional drained triaxial test is performed on a cylindrical sample of clay 50 mm in diameter and 100 mm in height. Young's modulus and Poisson's ratio for the clay are known to be 4.0 MPa and 0.25, respectively. A cell pressure of 100 kPa is applied to the sample. Following this, the deviatoric stress is increased from zero to 80 kPa. Calculate the diameter of the sample (i) after application of the cell pressure and (ii) following application of the deviatoric stress.

2.6 Suggest two ways to measure volumetric strain of a soil sample in a conventional drained triaxial compression test.

2.7 An unconventional triaxial compression test is performed on a sample of soft clay. The cell pressure and deviatoric stress are simultaneously altered through a servo-controlled system in such a way that the quantity $(\sigma_1 - \sigma_3)/(\sigma_1 + \sigma_3)$ is held constant. Assuming the clay is isotropic and elastic, derive expressions for the first and second invariants of the strain matrix $\boldsymbol{\epsilon}$, defined by

$$J_1 = tr\boldsymbol{\epsilon}$$

$$J_2 = \frac{1}{2}[(tr\boldsymbol{\epsilon})^2 - tr\boldsymbol{\epsilon}^2].$$

Derive an expression for the ratio $J_1/\sqrt{J_2}$, which involves only Poisson's ratio and the constant value of the quantity $(\sigma_1 - \sigma_3)/(\sigma_1 + \sigma_3)$.

2.8 An unlined conduit of diameter 1.0 m is located at a depth of 15 m below the surface of a deep deposit of soft sandstone as shown in the accompanying figure. The stress field around the conduit can be approximated by a radially symmetric stress state

$$\sigma_{rr} = \frac{p_o b^2}{(b^2 - a^2)} \left[1 - \frac{a^2}{r^2} \right]$$

$$\sigma_{\theta\theta} = \frac{p_o b^2}{(b^2 - a^2)} \left[1 + \frac{a^2}{r^2} \right]$$

$$\sigma_{r\theta} = \sigma_{\theta z} = \sigma_{rz} = 0$$

$$\sigma_{zz} = \nu(\sigma_{rr} + \sigma_{\theta\theta})$$

where $\nu = 0.25$ is Poisson's ratio, $a = 0.5$ m is the conduit radius, and $b = 5.0$ m is the radius of the zone of influence of the conduit. The in-situ stress p_o is 250 kPa. The soft sandstone has an elastic-brittle failure stress-strain behavior that obeys

$$(\tau_{\text{oct}})_{\text{max}} = 900 \text{ kPa}$$

where τ_{oct} is the octahedral shear stress (see question 1.6). The surface of the sandstone is to be uniformly loaded by a gravel fill with unit weight $\gamma = 20$ kN/m^3. Assuming that a fill of height H will increase the in-situ stress from p_o to $(p_o + \gamma H)$, calculate the maximum height of fill that can be placed to ensure there is a factor of safety of 3 against collapse of the conduit.

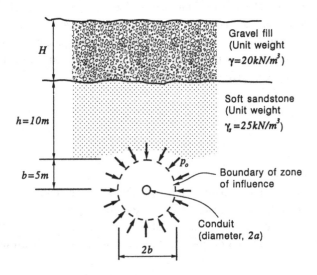

References

The works cited in Section 2.10 are:

Ohta, Y., and Goto, N., "Empirical shear wave velocity equations in terms of characteristic soil indexes," *Earthq. Engg Struct. Dyns.*, Vol. 6, pp. 167–187 (1978).

Parry, R. H. G., "Estimating foundation settlements in sand from plate bearing tests," *Geotechnique*, Vol. 28, pp. 107–118 (1978).

Schmertmann, J. H., "Static cone to compute static settlement over sand," *Jour. Soil Mech. Found. Div. ASCE*, SM3, pp. 1011–1043 (1970).

Selvadurai, A. P. S., and Nicholas, T. J., "A theoretical assessment of the screw plate test," *Proc. Third Int. Conf. Num. Methods Geomech.*, ed. W. Wittke, Aachen, Vol. 3, pp. 1245–1252 (1979).

Developments of generalized Hooke's law can be found in the elasticity references cited at the end of Chapter 1. The discovery that two constants were necessary to describe an isotropic material has an interesting history. It is described in

Timoshenko, S. P., *History of Strength of Materials*, McGraw-Hill, New York (1953).

The early work on soil testing mentioned in Section 2.9 can be traced to:

Darwin, G. H., "On the horizontal thrust of a mass of sand," *Minutes Proc. Inst. Civil Engrs.*, Vol. 71, pp. 350–378 (1883).

Bell, A. L., "Lateral pressure and resistance of clay and the supporting power of clay foundations," *Minutes Proc. Inst. Civil Engrs.*, Vol. 199, pp. 233–272 (1915).

3

Fundamental solutions

3.1 Boussinesq's problem

In this chapter, we will study some special problems that hold a fundamental position in relation to elastic solutions in geomechanics. Most of these problems were solved in the latter part of the nineteenth century, and they were usually solved not for application to geotechnical pursuits, but simply to answer basic questions about elasticity and the behavior of elastic bodies. With one exception, they all involve a *point load*. This is a finite force applied at a point: a surface of zero area. The point load is a mathematical artifice because we cannot achieve it in reality. Intuitively, we expect one or more components of stress to become infinite at the point where the load is placed, and this is realized in the mathematical solutions. Because of the stress singularities, understanding point-load problems will involve limiting procedures, which, as we know, are a bit dubious in regard to soils. But this should not deter us, since, in Chapter 4, we will pursue applications where the stresses remain finite, and the applied loads are not point loads.

Of all the point-load problems, the most useful in geomechanics is the problem of a point load acting normal to the surface of an elastic half-space. This problem was solved by the French mathematician Joseph Boussinesq in 1878. The problem geometry is illustrated in Figure 3.1. The half-space fills the volume $z \geq 0$, and we assume it is homogeneous, isotropic, and elastic. The point load is applied at the origin of coordinates on the half-space surface. It has magnitude P. Cylindrical coordinates are most convenient here. Later, we will think of the half-space as a deep deposit of soil, and z will represent depth beneath the ground surface.

The boundary conditions for this problem are as follows. Everywhere on the surface $z = 0$, except at the origin $r = 0$, tractions are specified zero. At the origin, the stresses must equilibrate the applied load P. Finally, for any point in the half-space infinitely distant from the origin, the displacements must all vanish.

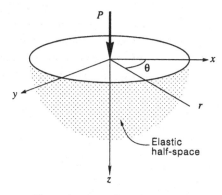

Figure 3.1 Boussinesq's problem.

If we were confronted with this problem, but did not happen to know Boussinesq had already solved it, we could make some guesses concerning its solution. One thing we would immediately guess is this: The solution must have *radial symmetry*. That means nothing can depend on the θ coordinate. Whatever angle θ we choose, the solution must look exactly the same. We could go even further and guess that the θ-component of the displacement vector u must be zero. We would do this because, if u_θ were not zero everywhere, there would be torsion of the half-space, at least at some points. It seems clear the point load could not cause any torsional motions. So, we expect to have a displacement vector of the form

$$u = \{u_r, 0, u_z\} \tag{3.1}$$

and moreover, u_r and u_z are not functions of θ.

Now our guessing has brought us to a general form for u. What does this imply about strains and stresses? We cannot directly use eqs. (1.2) and (1.3) here, because (1.2) was written for rectangular coordinates, and we now have a cylindrical coordinate system. In cylindrical coordinates the displacement gradient matrix will look like this

$$\nabla u = \begin{bmatrix} \dfrac{\partial u_r}{\partial r} & \dfrac{1}{r}\dfrac{\partial u_r}{\partial \theta} - \dfrac{u_\theta}{r} & \dfrac{\partial u_r}{\partial z} \\[2ex] \dfrac{\partial u_\theta}{\partial r} & \dfrac{1}{r}\dfrac{\partial u_\theta}{\partial \theta} + \dfrac{u_r}{r} & \dfrac{\partial u_\theta}{dz} \\[2ex] \dfrac{\partial u_z}{\partial r} & \dfrac{1}{r}\dfrac{\partial u_z}{\partial \theta} & \dfrac{\partial u_z}{\partial z} \end{bmatrix} \tag{3.2}$$

If u_θ is zero, and u_r and u_z cannot depend upon θ, then we see that ∇u will have only five nonzero components; the components lying on the two diagonals of the matrix. The strain-displacement relationship (1.3) then tells us that there will similarly be only five nonzero strain components, and finally if we use Hooke's law from (2.19) we see the stress matrix must have this form,

$$\boldsymbol{\sigma} = \begin{bmatrix} \sigma_{rr} & 0 & \sigma_{rz} \\ 0 & \sigma_{\theta\theta} & 0 \\ \sigma_{rz} & 0 & \sigma_{zz} \end{bmatrix}. \tag{3.3}$$

These stress components are illustrated in Figure 3.2. The three normal stresses are given particular names: σ_{zz} is the axial stress, σ_{rr} the radial stress, and $\sigma_{\theta\theta}$ is the hoop stress. Only one nonzero shear stress is present. (We would probably have guessed $\sigma_{r\theta}$ and $\sigma_{\theta z}$ would be zero just from the geometry of the point load, without first considering the displacements and strains.) Note in Figure 3.2 we have shown compressive stresses. We will adopt the compression positive sign convention in this chapter.

To proceed further we would need to study some solution techniques for elasticity problems, but that is not our object. It's more efficient from an engineering standpoint to just write out Boussinesq's solution. Boussinesq found the following displacement and stress fields for the vertical point-load problem

$$u_r = \frac{P}{4\pi GR}\left[\frac{rz}{R^2} - \frac{(1 - 2\nu)r}{R + z}\right]$$

$$u_\theta = 0 \tag{3.4}$$

$$u_z = \frac{P}{4\pi GR}\left[2(1 - \nu) + \frac{z^2}{R^2}\right]$$

and

$$\sigma_{rr} = -\frac{P}{2\pi}\left[\frac{(1 - 2\nu)}{R(R + z)} - \frac{3r^2 z}{R^5}\right]$$

$$\sigma_{\theta\theta} = -\frac{P(1 - 2\nu)}{2\pi}\left[\frac{z}{R^3} - \frac{1}{R(R + z)}\right]$$

$$\sigma_{zz} = \frac{P}{2\pi}\left[\frac{3z^3}{R^5}\right]$$

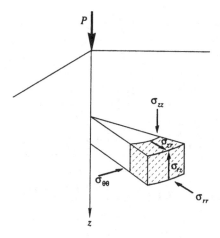

Figure 3.2 Boussinesq's stress field in cylindrical coordinates.

$$\sigma_{rz} = \frac{P}{2\pi}\left[\frac{3rz^2}{R^5}\right] = \sigma_{zr}$$

$$\sigma_{r\theta} = \sigma_{\theta r} = \sigma_{\theta z} = \sigma_{z\theta} = 0. \tag{3.5}$$

In all of these equations,

$$R^2 = r^2 + z^2. \tag{3.6}$$

For any point in the half-space, R is the straight-line distance to the origin.

Consider the stress field in eq. (3.5). We note that as R becomes large, all the stress components approach zero. On the boundary $z = 0$, we see that σ_{zz} and σ_{rz} vanish at every point, except at the origin of coordinates. At the origin, the stresses become singular, just as we expect for a point load. The unit normal vector to the boundary is $\hat{n} = [0, 0, -1]$. If we use this with eq. (1.12), we find that the traction vector on the boundary is zero everywhere except at the origin. We'll see how the singular stresses exactly equilibrate the point load in a moment. Figure 3.3 shows graphs of the distribution of σ_{zz} for various depths beneath the half-space surface. Note how the stress spreads laterally and diminishes in magnitude as depth increases.

Next, consider the displacement field. If we let R become large, both u_r and u_z approach zero, as our boundary condition specified. On the boundary $z = 0$ we have

$$u_r = -\frac{P(1 - 2\nu)}{4\pi Gr}, \quad u_z = \frac{P(1 - \nu)}{2\pi Gr} \tag{3.7}$$

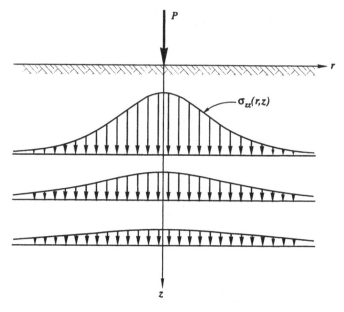

Figure 3.3 Distribution of the vertical stress σ_{zz} beneath the point load.

showing that both displacements become singular at the origin. This is an-other result of the point-load formalism. It should not create any anxiety since we won't deal with point loads later in applications.

Now, return to the stresses and find out how the stress field equilibrates the point load. Consider the hemispherical surface of radius a illustrated in Fig-ure 3.4. For any point on this surface we have $R = a = $ constant. Let ψ be the angle between a radius of the hemisphere and the z-axis. Then the unit nor-mal vector to the surface at any point can be written

$$\hat{n} = \begin{bmatrix} \sin\psi \\ 0 \\ \cos\psi \end{bmatrix} \tag{3.8}$$

while r and z components of the point are

$$z = a \cos\psi, \quad r = a \sin\psi. \tag{3.9}$$

The components of stress at any point on the surface are found by setting $R = a$ and using (3.9) in (3.5)

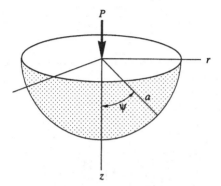

Figure 3.4 Hemispherical surface with radius a centered on the point of application of the point load.

$$\sigma_{rr} = -\frac{P}{2\pi}\left[\frac{1-2\nu}{a^2(1+\cos\psi)} - \frac{3\sin^2\psi\cos\psi}{a^2}\right]$$

$$\sigma_{\theta\theta} = -\frac{P(1-2\nu)}{2\pi}\left[\frac{\cos\psi}{a^2} - \frac{1}{a^2(1+\cos\psi)}\right]$$

$$\sigma_{zz} = \frac{3P\cos^3\psi}{2\pi a^2}$$ (3.10)

$$\sigma_{rz} = \frac{3P\sin\psi\cos^2\psi}{2\pi a^2}.$$

If we use (3.8) in (1.12), we can find the traction vector that acts on the hemispherical surface,

$$T = \begin{bmatrix} T_r \\ T_\theta \\ T_z \end{bmatrix} = \begin{bmatrix} \sigma_{rr} & 0 & \sigma_{rz} \\ 0 & \sigma_{\theta\theta} & 0 \\ \sigma_{rz} & 0 & \sigma_{zz} \end{bmatrix}\begin{bmatrix} \sin\psi \\ 0 \\ \cos\psi \end{bmatrix} = \begin{bmatrix} \sigma_{rr}\sin\psi + \sigma_{rz}\cos\psi \\ 0 \\ \sigma_{rz}\sin\psi + \sigma_{zz}\cos\psi \end{bmatrix}. \quad (3.11)$$

We could use (3.10) here to completely specify T in terms of P, a, and ψ.

Now think for a moment about the overall equilibrium of that part of the half-space contained inside the hemispherical surface. The horizontal upper surface just supports the vertical force P. The hemispherical surface is acted on by the tractions given in eq. (3.11). These tractions have one horizontal component T_r and one vertical component T_z. The horizontal components exactly cancel. For every T_r there is an equal and opposite component on the opposing side of the hemisphere.

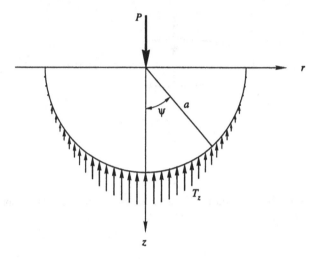

Figure 3.5 Vertical tractions acting on the hemispherical surface.

The vertical components T_z do not cancel. They have to combine to equilibrate the applied load P, as shown in Figure 3.5. We can write out T_z by using the last two equations of (3.10) in the appropriate places in (3.11)

$$T_z = \frac{3P}{2\pi a^2}\,[\sin^2\!\psi\,\cos^2\!\psi + \cos^4\!\psi] = \frac{3P}{2\pi a^2}\,\cos^2\!\psi. \qquad (3.12)$$

If we integrate T_z over the hemispherical surface, we will find the resultant upward force, which we expect to be P.

To integrate T_z, consider the elemental strip around the hemispherical surface shown in Figure 3.6. The strip width is $ad\psi$ and its total length is $2\pi a$ $\sin\psi$. The total upward force acting on the strip is found using (3.12)

$$T_z \times \text{Strip Area} = T_z \times 2\pi a^2 \sin\psi\,d\psi$$

$$= 3P\,\sin\psi\,\cos^2\!\psi\,d\psi$$

and we see something interesting here. The upward force on the strip depends upon ψ and P, but not on a. If we hold P and ψ constant, then this strip force will be the same on every hemispherical surface no matter how big or how small. Finally, we can find the resultant upward force that acts on the hemispherical surface by integrating

$$\text{Resultant upward force} = \int_0^{\pi/2} 3P\,\sin\psi\,\cos^2\!\psi\,d\psi = P \qquad (3.13)$$

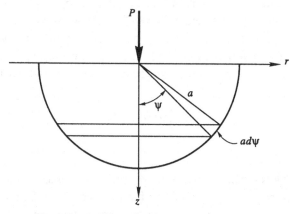

Figure 3.6 Geometry for integrating to find resultant force on hemispherical surface.

which is exactly what we expected to find. The interesting thing here is that this conclusion is true for any radius a. Even if we let a approach zero, and hence have infinite stresses, the resultant force will exactly equilibrate P.

Now we can almost state with confidence that Boussinesq's problem really is solved by (3.4) and (3.5). We have found that all the specified boundary conditions are satisfied. The last thing to verify is that the equilibrium equations are also satisfied. The form of the equilibrium equations we found in (1.23) will not be appropriate here since we are working in a cylindrical coordinate system. For cylindrical coordinates, eqs. (1.23) are replaced by

$$\frac{\partial \sigma_{rr}}{\partial r} + \frac{1}{r}\frac{\partial \sigma_{r\theta}}{\partial \theta} + \frac{\partial \sigma_{rz}}{\partial z} + \frac{1}{r}(\sigma_{rr} - \sigma_{\theta\theta}) + f_r = 0$$

$$\frac{\partial \sigma_{r\theta}}{\partial r} + \frac{1}{r}\frac{\partial \sigma_{\theta\theta}}{\partial \theta} + \frac{\partial \sigma_{\theta z}}{\partial z} + \frac{2}{r}\sigma_{r\theta} + f_\theta = 0 \qquad (3.14)$$

$$\frac{\partial \sigma_{rz}}{\partial r} + \frac{1}{r}\frac{\partial \sigma_{\theta z}}{\partial \theta} + \frac{\partial \sigma_{zz}}{\partial z} + \frac{1}{r}\sigma_{rz} + f_z = 0.$$

If we substitute for the stresses from eqs. (3.5), we will find these equations are exactly satisfied for the case where the body forces f_r, f_θ, and f_z are all zero.

One final point. To make certain of Boussinesq's solution, we should also show that the displacements in (3.4) are compatible with the stresses in (3.5) by finding the strains and using Hooke's law. We can find the strains by using (1.3) with the displacement gradient matrix defined by (3.2), and we can use these strains in Hooke's law to find the stresses given in (3.5). This is a straightforward but tedious business, and we will agree to accept that it can be done. An important thing to note here, if one should happen to feel ambitious and carry out these calculations; we need to treat signs carefully. If compressive stresses are taken positive, as in (3.5), the compressive strains must also be made positive. This necessitates a sign change in eq. (1.3). We require

$$\boldsymbol{\epsilon} = -\frac{1}{2} [\nabla \boldsymbol{u} + (\nabla \boldsymbol{u})^T]$$

in place of (1.3).

In Chapter 4, we will use Boussinesq's solution again and again to obtain both the displacement and stress fields for a number of applications in geotechnical engineering. Before we turn to applications, however, there are some other fundamental solutions that may not be quite so useful as Boussinesq's, but which need some consideration.

3.2 Flamant's problem

This is one of two non-point load problems we will take up in this chapter. Instead of a point load, we will consider a *line load* acting on the surface of a half-space as shown in Figure 3.7. The line load is similar to the point load in that we have a finite force acting on a surface of zero area, but now the load acts on an infinitely long line rather than on a point. We will represent this line load by $\wp$, and note that it has dimensions of force per unit length. The problem was solved by another French engineer, Alfred Flamant. Flamant was a colleague of Boussinesq and both were students of Saint-Venant. Flamant used Boussinesq's solution, together with the principal of superposition to solve for the stress field in the half-space. We will see how to accomplish this.

Consider the point marked A in Figure 3.7. The fact that this point lies beneath the x-axis doesn't mean any loss of generality. Since the load extends to infinity in both directions on the y-axis, we can place the origin of coordinates wherever we like on the y-axis. We'll try to find the σ_{zz} component of stress at point A due to the action of the line load. Figure 3.8 shows how to proceed. We can consider an elemental length dy of the y-axis. For any point inside the half-space, including point A, the line load $\wp$ acting on the element dy looks exactly like a point load. There will be an increment of stress $d\sigma_{zz}$ caused by this "point load" $\wp dy$. Boussinesq's solution tells us what $d\sigma_{zz}$ will be

$$d\sigma_{zz} = \frac{3(\wp \, dy)z^3}{2\pi R^5}. \tag{3.15}$$

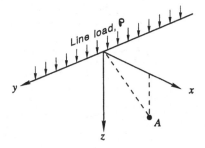

Figure 3.7 Flamant's problem.

This follows from the third equation of (3.5) with P replaced by $\wp\, dy$. The distances z and R are shown on Figure 3.8.

Now, to find the stress σ_{zz}, we can integrate both sides of (3.15)

$$\sigma_{zz} = \int_{-\infty}^{\infty} \frac{3\wp z^3}{2\pi R^5}\, dy. \tag{3.16}$$

To help with the integration we can introduce the angle ϕ shown in Figure 3.8. Letting $b = (x^2 + z^2)^{1/2}$ we have $y = b\tan\phi$ and $dy = b\sec^2\phi\, d\phi$. We can rewrite (3.16) as

$$\sigma_{zz} = \int_{-\pi/2}^{\pi/2} \frac{3\wp z^3}{2\pi b^4} \cos^3\phi\, d\phi. \tag{3.17}$$

Note that z and b are constants. Now we can easily integrate (3.17) to find

$$\sigma_{zz} = \frac{2\wp z^3}{\pi b^4} = \frac{2\wp z^3}{\pi(x^2 + z^2)^2}. \tag{3.18}$$

We can find the other components of stress by similar methods

$$\sigma_{xx} = \frac{2\wp x^2 z}{\pi(x^2 + z^2)^2}$$

$$\sigma_{yy} = \frac{2\wp \nu z}{\pi(x^2 + z^2)}$$

$$\sigma_{xz} = \frac{2\wp xz^2}{\pi(x^2 + z^2)^2} = \sigma_{zx}$$

$$\sigma_{xy} = \sigma_{yx} = \sigma_{zy} = \sigma_{yz} = 0.$$

$$\tag{3.19}$$

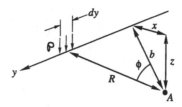

Figure 3.8 Geometry for integrating the point load to solve Flamant's problem.

(In fact, finding these components is not quite so easy as finding σ_{zz}. The vertical stress σ_{zz} doesn't differentiate between cylindrical and rectangular coordinates, but the other stress components do, and care must be taken.)

There are some interesting points here. One is that the line load is effectively a sequence of point loads side by side, and when we integrated (3.15) to get (3.16) we were using superposition. Had we not been dealing with linear elastic theory this might not have been possible. In one sense, Flamant's solution is just one of many applications of Boussinesq's solution, but it is a useful way to illustrate how we will approach many problems in Chapter 4. Another interesting point is that this is an example of a *plane-strain* problem. Plane-strain problems all have one spatial direction in which, at most, only rigid motions occur. As a result, certain strains will be identically zero. In Flamant's problem, we have $\epsilon_{yy} = \epsilon_{xy} = \epsilon_{yx} = \epsilon_{yz} = \epsilon_{zy} = 0$. Also, the nonzero strains are not functions of y. A particle that initially has coordinate y_0 in the reference configuration will always have coordinate y_0 in any deformed configuration unless a rigid translation in the y-direction occurs. It cannot move from the plane $y = y_0$. Plane-strain problems have many applications in geotechnical engineering.

It is also interesting to consider a cylindrical surface of radius b aligned with the line load as shown in Figure 3.9. We could carry out an analysis to find the tractions that act on this surface by using the stress components in (3.18) and (3.19). If we do so we find the traction vector is given by

$$T = \frac{2\wp z}{\pi b^2} \hat{n} \tag{3.20}$$

where $\hat{n}$ is the unit normal to the cylindrical surface. What this means is that the cylindrical surface itself is a principal surface. The major principal stress acts on it

$$\sigma_1 = \frac{2\wp z}{\pi b^2}. \tag{3.21}$$

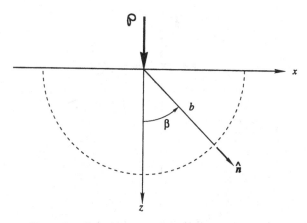

Figure 3.9 Cylindrical surface aligned with line load.

The intermediate principal surface is defined by $\hat{n} = \{0, 1, 0\}$ and the intermediate principal stress is $\sigma_2 = \nu\sigma_1$. The minor principal surface is perpendicular to the cylindrical surface and to the intermediate principal surface, and the minor principal stress is exactly zero.

It is a simple matter to integrate the vertical component of the traction (3.20) around the cylindrical surface in order to show that the resultant upward force exactly equilibrates the applied load $\wp$.

Another interesting characteristic of Flamant's problem is the distribution of the principal stress in space. Consider the locus of points on which the major principal stress σ_1 is a constant. From (3.21) we see this will be a surface for which

$$\frac{z}{b^2} = \frac{\pi\sigma_1}{2\wp} = \frac{1}{2c}$$

where c is a constant. This implies

$$b^2 = x^2 + z^2 = 2cz$$

and this is the equation of a circle with radius c centered on the z-axis at a depth c beneath the origin, as shown in Figure 3.10. At each point on this circle, the major principal stress is the same. It points directly at the origin. If we were to consider a larger value of c and consequently a larger circle, the value of σ_1 would be smaller in inverse proportion to c. This result gives rise to the idea of a *pressure bulb* in the soil beneath a foundation. It is a fairly vague idea at best, but may be helpful in visualizing stress fields.

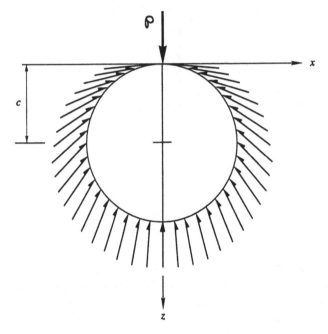

Figure 3.10 Locus of points on which the principal stresses are constant.

3.3 Kelvin's problem

Now we will return to another point-load problem. This is the problem of a point load acting in the interior of an infinite elastic body, shown in Figure 3.11. Rather than P, we will let the magnitude of the load be $2P$. This will help later when we compare the result of this problem with Boussinesq's solution. The problem was solved by the Scottish physicist William Thompson (who later became Lord Kelvin) in 1848.

Kelvin found the following solution, written in cylindrical coordinates

$$u_r = \frac{Prz}{8\pi G(1-\nu)R^3}$$

$$u_\theta = 0 \qquad\qquad\qquad (3.22)$$

$$u_z = \frac{P}{8\pi G(1-\nu)}\left[\frac{2(1-2\nu)}{R} + \frac{1}{R} + \frac{z^2}{R^3}\right]$$

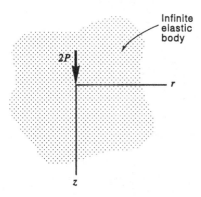

Figure 3.11 Kelvin's problem.

and

$$\sigma_{rr} = -\frac{P}{4\pi(1-\nu)}\left[\frac{(1-2\nu)z}{R^3} - \frac{3r^2z}{R^5}\right]$$

$$\sigma_{\theta\theta} = -\frac{P(1-2\nu)z}{4\pi(1-\nu)R^3}$$

$$\sigma_{zz} = \frac{P}{4\pi(1-\nu)}\left[\frac{(1-2\nu)z}{R^3} + \frac{3z^3}{R^5}\right] \tag{3.23}$$

$$\sigma_{rz} = \frac{P}{4\pi(1-\nu)}\left[\frac{(1-2\nu)r}{R^3} + \frac{3rz^2}{R^5}\right] = \sigma_{zr}$$

$$\sigma_{r\theta} = \sigma_{\theta r} = \sigma_{\theta z} = \sigma_{z\theta} = 0.$$

Here $R = \sqrt{z^2 + r^2}$, exactly as in the Boussinesq problem. If we examine the solution, we find singularities at the origin, where the point load acts, and we also find both displacements and stresses die out for large R. These are things we would expect. Note that on the plane $z = 0$, all of the stress components except for σ_{rz} vanish, at all points except the origin.

We can further investigate the solution by considering a planar surface defined by $z = c$, shown in Figure 3.12. The vertical component of traction on this surface is σ_{zz}. If we were to integrate σ_{zz} over this entire surface, what would be the resultant force? To find the answer to this question, note that the value of σ_{zz} will be a constant on any horizontal circle centered on the z-axis. Therefore, the force acting on the annulus shown in Figure 3.13 will be σ_{zz}

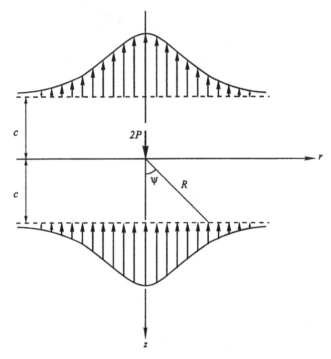

Figure 3.12 Distribution of vertical stress on horizontal planes above and below point load.

multiplied by the area $2\pi r dr$. The total resultant force on the surface $z = c$ is given by

$$\text{Resultant upward force} = \int_0^\infty \sigma_{zz}(2\pi r dr)$$

$$= \int_0^\infty \frac{P}{2(1 - \nu)} \left[\frac{(1 - 2\nu)c}{R^3} + \frac{3c^3}{R^5} \right] r dr.$$

To help integrate this, we can introduce the angle ψ shown in Figure 3.12. We have $r = c \tan\psi$ and $dr = c \sec^2\psi\, d\psi$. The integral becomes

$$\text{Resultant upward force} = \int_0^{\pi/2} \frac{P}{2(1 - \nu)} \left[(1 - 2\nu)\sin\psi + 3 \cos^2\psi \sin\psi \right] d\psi.$$

And when we evaluate this, we find the resultant force equals P, exactly one-half the applied load. This is not surprising since, if we were to consider

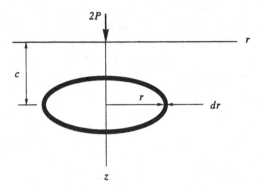

Figure 3.13 Geometry for integrating vertical stress shown in Figure 3.12.

a similar surface $z = -c$, also shown in Figure 3.12, we would find tensile stresses of the same magnitude as the compressive stresses on the lower plane. The resultant force on the upper plane would be $-P$, a tensile force. Together, the two resultant forces exactly equilibrate the applied load.

Finally, we can return to the earlier observation that, on the plane $z = 0$, all the stresses vanish except σ_{rz}. If we consider the special case where $\nu = 1/2$ (an incompressible material), then σ_{rz} will also be zero on this surface, and that part of the body below the $z = 0$ plane becomes equivalent to the half-space of Boussinesq's problem. Comparing Kelvin's solution (with $\nu = 1/2$) with Boussinesq's solution (with $\nu = 1/2$), we see they are identical for all $z \geq 0$. For $z \leq 0$ we also have Boussinesq's solution, but with a negative load $-P$. The two half-spaces, which together comprise the infinite body of Kelvin's problem, act as if they are uncoupled on the plane $z = 0$ where they meet. We find two Boussinesq problems wedded on the surface $z = 0$. If we consider a spherical surface centered on the origin we find it is a principal surface, supporting the major principal stress

$$\sigma_1 = \frac{3Pz}{2\pi R^3}$$

where R is the sphere radius. Note how σ_1 changes sign for negative values of z, giving tensile stresses above the median plane $z = 0$.

3.4 Cerrutti's problem

In 1882, the Italian, V. Cerrutti, solved the problem of a horizontal point load acting at the surface of an elastic half-space. Boussinesq also solved this problem, but evidently did so after Cerrutti without knowledge of Cerrutti's solution. The problem is illustrated in Figure 3.14. The point load is represented by P and acts at the origin of coordinates, pointing in the x-direction. This is a more complicated problem than either Boussinesq's or Kelvin's problem due to the absence of radial symmetry enjoyed by those two problems. We use a rectangular coordinate system.

Cerrutti's solution is

$$u_x = \frac{P}{4\pi GR}\left\{1 + \frac{x^2}{R^2} + (1 - 2\nu)\left[\frac{R}{R+z} - \frac{x^2}{(R+z)^2}\right]\right\}$$

$$u_y = \frac{P}{4\pi GR}\left\{\frac{xy}{R^2} - (1 - 2\nu)\frac{xy}{(R+z)^2}\right\} \qquad (3.24)$$

$$u_z = \frac{P}{4\pi GR}\left\{\frac{xz}{R^2} + (1 - 2\nu)\frac{x}{R+z}\right\}$$

and

$$\sigma_{xx} = -\frac{Px}{2\pi R^3}\left\{-\frac{3x^2}{R^2} + \frac{1-2\nu}{(R+z)^2}\left[R^2 - y^2 - \frac{2Ry^2}{R+z}\right]\right\}$$

$$\sigma_{yy} = -\frac{Px}{2\pi R^3}\left\{-\frac{3y^2}{R^2} + \frac{1-2\nu}{(R+z)^2}\left[3R^2 - x^2 - \frac{2Rx^2}{R+z}\right]\right\}$$

$$\sigma_{zz} = \frac{3Pxz^2}{2\pi R^5}$$

$$\sigma_{xy} = -\frac{Py}{2\pi R^3}\left\{-\frac{3x^2}{R^2} + \frac{1-2\nu}{(R+z)^2}\left[-R^2 + x^2 + \frac{2Rx^2}{R+z}\right]\right\} \qquad (3.25)$$

$$\sigma_{yz} = \frac{3Pxyz}{2\pi R^5}$$

$$\sigma_{zx} = \frac{3Px^2z}{2\pi R^5}.$$

Here

$$R^2 = x^2 + y^2 + z^2.$$

Inspecting the solution we find the expected singular displacements and stresses at the origin, while for large R, everything approaches zero. The vertical plane that contains the x-axis is a plane of symmetry. Looking at the x-component of the displacement field, we see that particles are displaced in the direction of the

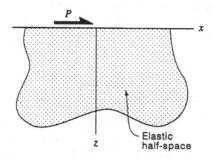

Figure 3.14 Cerrutti's problem.

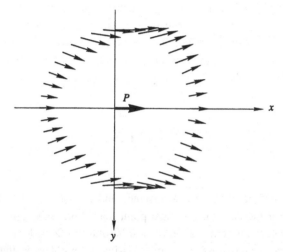

Figure 3.15 Distribution of horizontal displacements on a circle surrounding a horizontal point load.

point load. The y-component of displacement moves particles away from the x-axis for positive values of x and toward the x-axis for negative x. A plot of horizontal displacement vectors at the surface $z = 0$ is shown in Figure 3.15 for the special case of an incompressible material. Vertical displacements take the sign of x and hence particles move downward in front of the load and upward behind the load. Both the displacements and stresses for Cerrutti's problem appear more complex than for either Boussinesq's or Kelvin's problem, but this is largely an illusion. If we were to recast Boussinesq's or Kelvin's solutions in rectangular coordinates, they would look much the same as Cerrutti's solution. In fact, Cerrutti's solution is embedded in Kelvin's solution for the special case where $\nu = 1/2$. If we set ν equal to 1/2 in both solutions we find the stresses and displacements in the left- or right-hand half-space of Kelvin's problem are identical to Cerrutti's solution, provided we designate the coordinate axes appropriately.

3.5 Mindlin's problem

The final two variations of the point-load problem that we will consider in detail were solved by Raymond Mindlin in 1936. These are the problems of a point load (either vertical or horizontal) acting in the interior of an elastic half-space. Mindlin's problem is illustrated in Figure 3.16. The load P acts at a point located a distance z beneath the half-space surface. This problem is more complex than Boussinesq's or Kelvin's or Cerrutti's. It has found application in considerations of the stress and displacement fields surrounding an axially loaded pile and in the study of interaction between foundations and ground anchors.

It is useful, in writing out Mindlin's solution, to place the origin of coordinates a distance c above the free surface, as shown in Figure 3.16. Then the applied load acts at the point $z = 2c$. Also, we introduce the definitions

$$z_1 = z - 2c$$
$$R^2 = r^2 + z^2 \qquad (3.26)$$
$$R_1^2 = r^2 + z_1^2.$$

Thus z_1 and R_1 are the vertical distance and the radial distance from the point load. For the case of a vertical point load, Mindlin's solution is most conveniently stated in terms of Boussinesq's solution. Consider for a moment the displacement and stress fields in Boussinesq's problem in the region of the half-space below the surface $z = c$. These displacements and stresses are also found in Mindlin's solution, but with additional terms. The equations written out below give these additional terms. To obtain the complete solution we must add them to eqs. (3.4) and (3.5)

$$u_r = \frac{Pr}{16\pi(1-\nu)G} \left\{ \frac{z_1}{R_1^3} - \frac{z - 2(3-4\nu)c}{R^3} + \frac{6cz(z-c)}{R^5} \right\}$$

$$u_\theta = 0$$

$$u_z = \frac{P}{16\pi(1-\nu)G} \left\{ \frac{z_1^2}{R_1^3} + \frac{3-4\nu}{R_1} - \frac{3-4\nu}{R} \right. \qquad (3.27)$$

$$\left. - \frac{z^2 + 2cz - 2c^2}{R^3} + \frac{6cz^2(z-c)}{R^5} \right\}$$

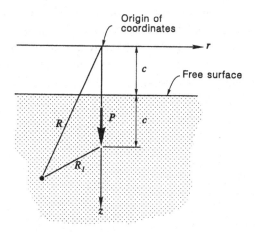

Figure 3.16 Mindlin's problem.

$$\sigma_{rr} = \frac{P}{8\pi(1-\nu)} \left\{ \frac{3r^2 z_1}{R_1^5} - \frac{(1-2\nu)z_1}{R_1^3} + \frac{(1-2\nu)z - 12(1-\nu)c}{R^3} \right.$$

$$\left. - \frac{3r^2 z - 6(7-2\nu)cz^2 + 24c^2 z}{R^5} - \frac{30cz^2(z-c)}{R^7} \right\}$$

$$\sigma_{\theta\theta} = \frac{P}{8\pi(1-\nu)} \left\{ - \frac{(1-2\nu)z_1}{R_1^3} + \frac{(1-2\nu)(z+6c)}{R^3} \right.$$

$$\left. - \frac{6(1-2\nu)cz^2 - 6c^2 z}{R^5} \right\}$$

$$\sigma_{zz} = \frac{P}{8\pi(1-\nu)} \left\{ \frac{3z_1^3}{R_1^5} + \frac{(1-2\nu)z_1}{R_1^3} - \frac{(1-2\nu)(z-2c)}{R^3} \right. \qquad (3.28)$$

$$\left. - \frac{3z^3 + 12(2-\nu)cz^2 - 18c^2 z}{R^5} + \frac{30cz^2(z-c)}{R^7} \right\}$$

$$\sigma_{rz} = \frac{Pr}{8\pi(1-\nu)} \left\{ \frac{3z_1^2}{R_1^5} + \frac{1-2\nu}{R_1^3} - \frac{1-2\nu}{R^3} \right.$$

$$\left. - \frac{3z^2 + 6(3-2\nu)cz - 6c^2}{R^5} + \frac{30cz^2(z-c)}{R^7} \right\} = \sigma_{zr}$$

$$\sigma_{r\theta} = \sigma_{\theta r} = \sigma_{\theta z} = \sigma_{z\theta} = 0.$$

Mindlin's solution for a horizontal point load also employs the definitions for z_1 and R_1, but now we must introduce rectangular coordinates because there is no longer cylindrical symmetry. We replace r^2 by $x^2 + y^2$, and assume (without any loss of generality) that the load acts in the x-direction at the point $z = c$. The solution here is conveniently stated in terms of Cerrutti's solution, just as the vertical point load was given in terms of Boussinesq's solution. Thus, the displacements and stresses to be superposed on Cerrutti's solution are

$$u_x = \frac{P}{16\pi(1-\nu)G}\left\{\frac{x^2}{R_1^3} + \frac{3-4\nu}{R_1} - \frac{3-4\nu}{R}\right.$$
$$\left. + \frac{-x^2 + 2c(z-c)}{R^3} - \frac{6cx^2(z-c)}{R^5}\right\}$$

$$u_y = \frac{P}{16\pi(1-\nu)G}\left\{\frac{xy}{R_1^3} - \frac{xy}{R^3} - \frac{6cxy(z-c)}{R^5}\right\}$$

$$u_z = \frac{P}{16\pi(1-\nu)G}\left\{\frac{xz_1}{R_1^3} - \frac{xz + 2(3-4\nu)cx}{R^3} - \frac{6cxz(z-c)}{R^5}\right\}$$

(3.29)

and

$$\sigma_{xx} = \frac{Px}{8\pi(1-\nu)}\left\{\frac{3x^2}{R_1^5} + \frac{1-2\nu}{R_1^3} - \frac{1-2\nu}{R^3}\right.$$
$$\left. - \frac{3x^2 - 6(3-2\nu)cz + 18c^2}{R^5} - \frac{30cx^2(z-c)}{R^7}\right\}$$

$$\sigma_{yy} = \frac{Px}{8\pi(1-\nu)}\left\{\frac{3y^2}{R_1^5} - \frac{1-2\nu}{R_1^3} + \frac{1-2\nu}{R^3}\right.$$
$$\left. - \frac{3y^2 - 6(1-2\nu)cz + 6c^2}{R^5} - \frac{30cy^2(z-c)}{R^7}\right\}$$

$$\sigma_{zz} = \frac{Px}{8\pi(1-\nu)}\left\{\frac{3z_1^2}{R_1^5} - \frac{1-2\nu}{R_1^3} + \frac{1-2\nu}{R^3}\right.$$
$$\left. - \frac{3z^2 + 6(1-2\nu)cz + 6c^2}{R^5} - \frac{30cz^2(z-c)}{R^7}\right\}$$

$$\sigma_{xy} = \frac{Py}{8\pi(1-\nu)}\left\{\frac{3x^2}{R_1^5} + \frac{1-2\nu}{R_1^3} - \frac{1-2\nu}{R^3}\right.$$
$$\left. - \frac{3x^2 - 6c(z-c)}{R^5} - \frac{30cx^2(z-c)}{R^7}\right\} = \sigma_{yx}$$

$$\sigma_{yz} = \frac{Pxy}{8\pi(1-\nu)}\left\{\frac{3z_1}{R_1^5} - \frac{3z + 6(1-2\nu)c}{R^5} - \frac{30cz(z-c)}{R^7}\right\} = \sigma_{zy}$$

$$\sigma_{xz} = \frac{P}{8\pi(1-\nu)}\left\{\frac{3x^2z_1}{R_1^5} + \frac{(1-2\nu)z_1}{R_1^3} - \frac{(1-2\nu)(z-2c)}{R^3}\right.$$ (3.30)

$$\left. - \frac{3x^2z + 6(1-2\nu)cx^2 - 6cz(z-c)}{R^5} - \frac{30cx^2z(z-c)}{R^7}\right\} = \sigma_{zx}.$$

Now consider the solution for the vertical load in eqs. (3.27) and (3.28). If we let the distance c approach infinity, then we should recover Kelvin's solution for the infinite body. We can see that $c \to \infty$ implies both z and R become infinite, and all terms in (3.27) and (3.28) that involve z and R disappear. Those terms that are left involve z_1 and R_1, but we realize z_1 and R_1 are exactly the same as z and R in Kelvin's solution. We also realize that eqs. (3.27) and (3.28) must be added to Boussinesq's solution to give the complete displacement and stress fields for Mindlin's problem, but, with $R \to \infty$, all the Boussinesq terms reduce to zero. Finally, we must remember that Kelvin's problem involved a point load of $2P$, while Mindlin's solution represents a point load P. Keeping these points in mind, we can see that (3.27) and (3.28) will reduce to Kelvin's solution when c approaches infinity. Exactly the same conclusion applies to Mindlin's second problem, but it is not quite so easy to see the result, since we must interchange the x and z axes and introduce cylindrical coordinates in (3.29) and (3.30).

Returning again to the vertical point load, we can immediately see from (3.27) and (3.28) that, if we set $c = 0$ (and consequently $z_1 = z$, $R_1 = R$), all of the displacements and stresses vanish. This is exactly what we expect, since Mindlin's problem with $c = 0$ is equivalent to Boussinesq's problem, and eqs. (3.27) and (3.28) are the *differences* from Boussinesq's solution. A similar conclusion applies to (3.29) and (3.30) with $c = 0$, taking account of Cerrutti's solution. Thus Mindlin's solutions properly correspond to Boussinesq's and Cerrutti's solutions for the special case of $c = 0$.

Finally, consider the free surface $z = c$. If we set z equal to c in (3.26), we find $z_1 = -c$ and $R = R_1$. On the free surface, we must have zero traction, implying for the case of cylindrical symmetry that $\sigma_{zz} = \sigma_{rz} = \sigma_{zr} = 0$. Setting $z = c$ in Mindlin's equations for σ_{zz} and σ_{rz} in (3.28), we find both stresses become equal to Boussinesq's σ_{zz} and σ_{rz} (with $z = c$), but with signs reversed. Thus, when we superpose eqs. (3.28) on Boussinesq's solution, to get the complete stress field for Mindlin's problem, we find the boundary $z = c$ is free from traction, as we expect. A similar comment applies regarding the stress components σ_{zz}, σ_{xz}, σ_{yz} in eqs. (3.30) when superposed on Cerrutti's solution.

3.6 Other fundamental solutions

In this section, we can make a few general notes on other point-load problems and their solutions. These problems are generally not of such great interest in geotechnical engineering, but they may be useful in special circumstances. The first of these additional problems is the case of an elastic layer resting on an elastic half-space with different elastic properties as in Figure 3.17. Two categories of this problem arise. In one, the interface between the layer and the half-space is smooth and different horizontal displacements may occur above and below the interface. In the second variation, the interface is rough, and displacements are the same in both materials. This problem was first solved by Donald Burmister in 1943 for the special case of incompressible materials (i.e., $\nu = 0.50$). In later papers, he extended his solution to cover compressible materials and multiple layers. At first glance, it may appear this solution is particularly useful in geomechanics, since in many practical situations we find a layer of soil resting on bedrock. Unfortunately, Burmister's solutions are complex and difficult to apply. Instead, a simple approximate solution for layered soils can be used. This is discussed in Chapter 4.

A second category of problems involves dynamic loads in contrast to the static problems we have considered so far. In these problems, the load is a function of time. The simplest problem is a dynamic point load acting in the interior of an infinite elastic body, the dynamic analogy to Kelvin's problem. It was solved by the Irish mathematician George Stokes in 1849, only one year after Kelvin had solved the static problem. Stokes' solution shows that body waves of both types, longitudinal and transverse, radiate outward from the point load. One other dynamic problem deserving mention is Lamb's problem. This is the dynamical equivalent of Flamant's problem with a dynamic line load on the surface of an elastic half-space. It was solved by Horace Lamb in 1904.

A third group of point-load problems are those involving anisotropic or continuously nonhomogeneous elastic half-spaces. The solution for a point load on the boundary of an anisotropic elastic half-space was determined by the Australian mathematician J. H. Michell in 1900. Michell considered a particular type of anisotropy called transverse isotropy. The elasticity of the body is different in horizontal and vertical directions, and the z-axis is an axis of elastic symmetry. For this case, five independent elastic constants

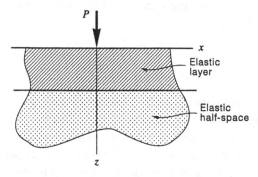

Figure 3.17 Point load on an elastic layer.

are required to fully characterize the body; three are needed to relate normal stresses to extensional strains, and there are two independent shear moduli. Problems involving nonhomogeneous elastic bodies were also solved by another Australian, John Booker, together with two coworkers, in 1985. These solutions apply specifically to elastic materials in which the value of Young's modulus increases with increasing depth beneath the free surface, while Poisson's ratio remains constant. Both point-load and line-load solutions were obtained.

Looking back over the array of fundamental solutions, it is probably clear that Boussinesq's will be the most useful to the geotechnical engineer. It has the great virtue of simplicity. Moreover, the two most interesting stress components, σ_{zz} and σ_{rz}, have solutions that do not depend on either of the elastic constants. Following the publication of Boussinesq's solution, engineers throughout the world began to make use of it, but a peculiar sequence of events occurred. It was an easy matter to integrate Boussinesq's solution over some regions of the half-space surface to obtain a solution for a distributed pressure. We will do exactly that at the beginning of Chapter 4. This was done by a number of people, including Boussinesq himself. The simplest loaded region to integrate was circular, and the σ_{zz} distribution in an elastic half-space subjected to a uniform circular load became well-known. Experiments were then performed and comparisons between measured and computed values of stress showed considerable differences. In fact, the differences resulted from two experimental difficulties: the inability to apply a *uniform* stress over the circular region and the use of a thin layer of soil to represent a half-space. Unfortunately, the differences were interpreted as a failure of Boussinesq's theory, and investigators set out to adjust the Boussinesq equations so that they would agree with the experiments. This was done by replacing the solution for σ_{zz} by this equation

$$\sigma_{zz} = \frac{nPz^n}{2\pi R^{n+2}}.$$

If we compare this with eqs. (3.5), we see Boussinesq's result corresponds to the case where $n = 3$. The parameter n was called the *concentration factor*, and a great deal of interest was focused on what value it should have for different types of soil conditions. None of this makes much sense, of course. A material in which the stress distribution is governed by this equation is inelastic for any value of n other than 3. And, for an inelastic body, the principle of superposition is inapplicable, hence integrating over a distributed load has no basis. Nevertheless, the idea of the concentration factor held on for many years. There is an adage applicable to circumstances like this. It contrasts theory with experiment, and loosely stated, reads "No one believes the theoretician, except the theoretician himself. Everyone believes the experimentalist, except for the experimentalist himself."

3.7 Gravity stresses, stress functions

To conclude this chapter, we can make some observations about the stress state in a soil deposit caused by the self weight of the soil. We will call these stresses the *gravity stresses*. They may be of interest in their own right; for example, when we are constructing an earth embankment or considering the stability of a natural slope. In other problems, they may be of secondary interest. For example, we may wish to know the value of some component of stress beneath a loaded footing. In general, the stress will be partly due to the footing load and partly due to the self weight of the soil. For this situation we can use superposition and add the two effects to obtain the total value of the stress. In the next chapter we will deal with the problem of the stress caused by the footing. Here we will only consider the self-weight problem.

To begin, consider a soil deposit with a horizontal surface (and possibly horizontal layering as in Figure 3.18). It seems reasonable to assume σ_{xx} and σ_{yy} are equal and also to assume the shear stresses σ_{xy}, σ_{yz}, and σ_{xz} will all be zero. Moreover, none of the stresses should be a function of x or y. The equilibrium equations (1.23) are identically satisfied except for the third equation, which reduces to

$$\frac{\partial \sigma_{zz}}{\partial z} + f_z = 0. \tag{3.31}$$

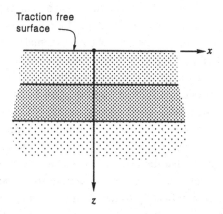

Figure 3.18 Gravity stresses: horizontal ground surface.

The body force f_z is $-\rho g$, the negative unit weight of the soil. Here $\rho = \rho(z)$ is the soil mass density, g is the acceleration of gravity, and the negative sign is used because we wish compressive stress to be positive. We can integrate (3.31) to have

$$\sigma_{zz}(z) = \int_0^z \rho(z)g \, dz. \tag{3.32}$$

Referring back to (3.31) we note that the partial derivative there would suggest that we should add an arbitrary function of x and y to the right-hand side of (3.32); but we have already agreed that there should be no dependence on either x or y, so that function has been dropped. For a homogeneous soil profile, (3.32) reduces to $\sigma_{zz} = \rho g z$.

Now what are the horizontal stresses σ_{xx} and σ_{yy}? The answer is, we do not know. If the soil is perfectly elastic, then we might argue that there can be no horizontal deformation in the half-space. Setting ϵ_{xx} and ϵ_{yy} equal to zero in Hooke's law [see equation (2.5)] we find

$$\sigma_{xx} = \sigma_{yy} = \frac{\nu}{1 - \nu}\sigma_{zz}. \tag{3.33}$$

But this result depends on our assumption of elastic behavior, and that may not be valid. In Section 2.9, we discussed the possibility of irreversible volumetric strains that may occur in natural deposits. This led to the concepts of normally- and over-consolidated soils. Any irreversible straining will, of course, result in inelastic behavior, and then we can no longer bank on Hooke's law with our assumption of zero horizontal deformation. Geotechnical engineers have made careful measurements* of the horizontal components of stress σ_{xx} and σ_{yy} in natural soil deposits. These measurements show that σ_{xx} and σ_{yy} may take on a wide range of values depending on the past history of the soil. In normally consolidated soils, σ_{xx} and σ_{yy} will be smaller than σ_{zz}, and (3.33) may give a reasonably good approximation for their value. In overconsolidated soils, however, σ_{xx} and σ_{yy} will usually be greater than σ_{zz}, something we could never achieve from (3.33) since ν can never exceed 0.5. We see that the question of what the gravity stresses are is not a simple one. Indeed, some pressuremeter measurements in certain soil deposits suggest that σ_{xx} and σ_{yy} may take significantly different values themselves.

For a soil deposit with horizontal surface, we can at least find the vertical stress σ_{zz}. What if the surface is not horizontal? Then the question of the gravity stresses becomes a bit more interesting. It also allows us to introduce a simple but effective means of finding solutions to a particular class of elasticity problem. The method involves the use of a tool called a *stress function*.

There are a number of stress functions used in the theory of elasticity, but the one we will be concerned with is the Airy stress function, first used by the British astronomer and mathematician George B. Airy in 1862. The Airy stress function is well-known and is described in nearly every elasticity textbook, so we will only briefly discuss it here.

The class of problem we are concerned with is plane strain. We discussed plane-strain conditions in regard to Flamant's problem in Section 3.2. Recall that for plane-strain problems there is one direction in which, at most, only rigid motions occur. Suppose we take this to be the y-direction. Then the strains ϵ_{yy}, ϵ_{xy}, ϵ_{yx}, ϵ_{yz}, and ϵ_{zy} will all be zero, and the nonzero strains ϵ_{xx}, ϵ_{zz}, and $\epsilon_{xz} = \epsilon_{zx}$ will not be functions of y. Hooke's law then shows that the only nonzero stresses are σ_{xx}, σ_{yy}, σ_{zz}, and $\sigma_{xz} = \sigma_{zx}$. The fact that ϵ_{yy} is zero also shows that $\sigma_{yy} = \nu(\sigma_{xx} + \sigma_{zz})$, so there are only three independent stress components to fully describe a plane-strain stress field: σ_{xx}, σ_{zz}, and σ_{xz}.

If we look to the equilibrium equations (1.23), we now see that for plane-strain conditions they reduce to two equations

$$\frac{\partial \sigma_{xx}}{\partial x} + \frac{\partial \sigma_{xz}}{\partial z} + f_x = 0$$

$$\frac{\partial \sigma_{xz}}{\partial x} + \frac{\partial \sigma_{zz}}{\partial z} + f_z = 0. \tag{3.34}$$

---

* These are usually made using a pressuremeter, which we discussed near the end of Chapter 2.

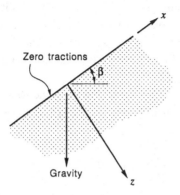

Figure 3.19 Gravity stresses: sloping ground surface.

For gravity stress problems, the body forces f_x and f_z must account for the self weight of the soil. We have already seen that with the geometry of Figure 3.18 the component f_z becomes $-\rho g$. The value of f_x, of course, is zero. For a slightly more general situation such as that shown in Figure 3.19, f_x and f_z are given by

$$f_x = \rho g\,\sin\beta \quad \text{and} \quad f_z = -\rho g\,\cos\beta. \qquad (3.35)$$

Here β represents the angle between the direction of gravity and the z-axis. A convenient way to represent the effects of gravity in the equilibrium equations is to introduce a potential function $V = V(x,z)$. If we let

$$V = \rho g(-x\,\sin\beta + z\,\cos\beta) \qquad (3.36)$$

then (3.34) can be written

$$\frac{\partial \sigma_{xx}}{\partial x} + \frac{\partial \sigma_{xz}}{\partial z} - \frac{\partial V}{\partial x} = 0$$

$$\frac{\partial \sigma_{xz}}{\partial x} + \frac{\partial \sigma_{zz}}{\partial z} - \frac{\partial V}{\partial z} = 0. \qquad (3.37)$$

This will be the form of the equilibrium equations for which we will seek solutions.

The other equations we need to worry about are the compatibility equations from Section 1.6. To see why this is so, consider a hypothetical situation in which we have managed to solve equations (3.37) for the stresses σ_{xx}, σ_{zz}, and σ_{xz} subject to some given set of boundary conditions. If we have the stresses, we can use Hooke's law to find the corresponding strains. But can we integrate the strains to find displacements? The answer of course is yes, provided the compatibility equations are satisfied, and that is why we need to consider equations (1.11).

Equations (1.11) look formidable, but for plane-strain conditions only one, the third, is not automatically satisfied. Rewriting it we have

$$\frac{\partial^2 \epsilon_{xx}}{\partial z^2} + \frac{\partial^2 \epsilon_{zz}}{\partial x^2} = 2 \frac{\partial^2 \epsilon_{xz}}{\partial x \partial z}. \tag{3.38}$$

We would like to recast this equation now in terms of stresses. For plane-strain conditions, Hooke's law, (2.5) and (2.6), can be written

$$2G\epsilon_{xx} = \sigma_{xx} - \nu(\sigma_{xx} + \sigma_{zz})$$

$$2G\epsilon_{zz} = \sigma_{zz} - \nu(\sigma_{xx} + \sigma_{zz}) \tag{3.39}$$

$$2G\epsilon_{xz} = \sigma_{xz}.$$

Using these in (3.38) gives

$$\frac{\partial^2 \sigma_{xx}}{\partial z^2} + \frac{\partial^2 \sigma_{zz}}{\partial x^2} - \nu\nabla^2(\sigma_{xx} + \sigma_{zz}) = 2 \frac{\partial^2 \sigma_{xz}}{\partial x \partial z}. \tag{3.40}$$

Here $\nabla^2 = \dfrac{\partial^2}{\partial x^2} + \dfrac{\partial^2}{\partial z^2}$ is the two-dimensional Laplacian operator.

Now if we differentiate the first of the equilibrium equations (3.37) with respect to x and the second with respect to z and add the resulting equations, we find

$$2 \frac{\partial^2 \sigma_{xz}}{\partial x \partial z} = - \frac{\partial^2 \sigma_{xx}}{\partial x^2} - \frac{\partial^2 \sigma_{zz}}{\partial z^2} + \nabla^2 V. \tag{3.41}$$

Using this in (3.40) gives the following expression for the compatibility equation we are concerned with

$$(1 - \nu)\nabla^2(\sigma_{xx} + \sigma_{zz}) - \nabla^2 V = 0. \tag{3.42}$$

Now we have three equations, (3.37) and (3.42), to solve for the three unknown stresses σ_{xx}, σ_{xz}, and σ_{zz}.

How might we find a solution? Airy suggested we consider a function $\Phi = \Phi(x, z)$ constructed in such a way that

$$\sigma_{xx} = \frac{\partial^2 \Phi}{\partial z^2} + V, \ \sigma_{zz} = \frac{\partial^2 \Phi}{\partial x^2} + V, \ \sigma_{xz} = -\frac{\partial^2 \Phi}{\partial x \partial z}. \tag{3.43}$$

The form of these equations ensures that the equilibrium equations (3.37) will be satisfied. And if we use the first two of (3.43) in (3.42) we find

$$\nabla^2 \left(\frac{\partial^2 \Phi}{\partial x^2} + \frac{\partial^2 \Phi}{\partial z^2} \right) + \frac{1 - 2\nu}{1 - \nu} \nabla^2 V = 0. \tag{3.44}$$

Now note from (3.36) that V is only linear in x and z, so that $\nabla^2 V$ will equal zero. Then (3.44) reduces to

$$\nabla^4 \Phi = 0 \tag{3.45}$$

where ∇^4, called the *biharmonic operator*, is defined by

$$\nabla^4 = \nabla^2(\nabla^2) = \frac{\partial^4}{\partial x^4} + 2\frac{\partial^4}{\partial x^2 \partial z^2} + \frac{\partial^4}{\partial z^4}. \tag{3.46}$$

Now it becomes clear how the stress function can help us to find solutions. Any function that satisfies the biharmonic equation (3.45) offers a possible solution. The stresses determined from (3.43) will satisfy both equilibrium and compatibility. If the stresses also satisfy the given boundary conditions, then we have a particular solution.

We can illustrate this by considering the gravity-stress problem where the ground surface is sloping, as shown in Figure 3.19. The situation in Figure 3.19 might represent part of a long natural slope. We are aware the ground surface cannot continue at a constant slope to ± infinity in the x-direction, but Saint-Venant's principle (Appendix E) tells us that any change in the ground profile that happens at some reasonable distance away is not likely to affect the stresses in the region of interest. Note that gravity acts at an angle β to the z-direction so that the gravity potential V is given by (3.36).

How should we proceed to solve for the gravity stresses in the sloping soil deposit? We will try to find an Airy stress function to do the job. Any function we propose must first satisfy the biharmonic equation (3.45) and then must produce stresses that satisfy the boundary conditions. The only boundary condition we have here is that the ground surface is traction-free. That is

$$\text{on } z = 0, \ \sigma_{zz} = \sigma_{xz} = 0 \qquad (3.47)$$

which must be true for all values of x.

At this point we might begin to consider any functions $\Phi(x,z)$, which are biharmonic. There are quite a number of functions that will automatically satisfy (3.45). The simplest are all polynomials in x and z of degree three or less. Suppose we try a polynomial such as

$$\Phi\ (x,z) = c_1 x^3 + c_2 x^2 z + c_3 x z^2 + c_4 z^3. \qquad (3.48)$$

Here the coefficients c_1, c_2, c_3, c_4 are constants, but otherwise are unspecified. It is obvious this function will satisfy (3.45), but what about the boundary conditions? If we use (3.48) in (3.43) we find the stresses are given by

$$\sigma_{xx} = 2c_3 x + 6c_4 z + \rho g(-x \sin\beta + z \cos\beta)$$

$$\sigma_{zz} = 6c_1 x + 2c_2 z + \rho g(-x \sin\beta + z \cos\beta) \qquad (3.49)$$

$$\sigma_{xz} = -2c_2 x - 2c_3 z.$$

We want these stresses to satisfy the boundary conditions. Setting $z = 0$ we see from (3.47) that we must have

$$c_2 = 0 \quad \text{and} \quad c_1 = \frac{1}{6} \rho g \sin\beta.$$

Using these in (3.49) we now have σ_{xx} unchanged and

$$\sigma_{zz} = \rho g z \cos\beta, \quad \sigma_{xz} = -2c_3 z. \qquad (3.50)$$

Can we proceed any farther? We can, but to do so we must say something about the lateral stress σ_{xx}. Just as in the case of the horizontal ground surface at the beginning of this section, we find σ_{xx} a bit perplexing here as well. One thing we can say about σ_{xx} is that its value should not depend upon x. As (3.49) makes clear, σ_{xx} may be linear in x, but this would imply that if x becomes very large then σ_{xx} could grow without bound. Since we would like to have finite stresses only, it's reasonable to require σ_{xx} not depend upon x. This implies

$$c_3 = \frac{1}{2} \rho g \sin\beta.$$

This is about as far as we can go without other information. The stresses are now given by

$$\sigma_{xx} = (6c_4 + \rho g \cos\beta)z$$

$$\sigma_{zz} = \rho g z \cos\beta \qquad (3.51)$$

$$\sigma_{xz} = -\rho g z \sin\beta.$$

The coefficient c_4 remains unknown. If we were to assume zero strain in the x-direction ($\epsilon_{xx} = 0$) as we did in the horizontal ground-surface problem, then (3.39) shows

$$\sigma_{xx} = \frac{\nu}{1 - \nu} \sigma_{zz}$$

just as in (3.33), and c_4 would be given by

$$c_4 = -\frac{1}{6} \left(\frac{1 - 2\nu}{1 - \nu} \right) \rho g \cos\beta.$$

But sloping ground is no different from horizontal ground and the questions about σ_{xx} are no more easily answered here than they were at the beginning of this section. The other two components of stress, σ_{zz} and σ_{xz}, are well-determined however, and we have learned a little about the Airy stress function, a useful component of the elastician's toolbox.

Exercises

3.1 Consider the hemispherical surface of radius a shown in Figure 3.4. Make a graph showing how σ_{zz} varies on this surface as a function of the angle ψ. Find the value of ψ for which σ_{zz} is maximum. Carry out the same operations for σ_{rr} and σ_{rz}. Take ν to be 0.25.

3.2   In the Boussinesq point-load problem, let u_R denote the displacement
      in the R-direction (where $R^2 = r^2 + z^2$). Then u_R is the displacement
      directed either toward or away from the point load. On the half-space
      surface where $z = 0$, $u_R = u_r$, and we see from eq. (3.4) that u_R is di-
      rected toward the point load. On the z-axis (where $r = 0$), $u_R = u_z$, and
      we see that u_R is directed away from the point load. Find the surface
      where u_R is exactly zero.

3.3   Consider the elastic half-space supporting two equal point loads, P, il-
      lustrated below. Using the principle of superposition, find an expression
      for the vertical displacement of the half-space surface $u_z(x, y, z = 0)$ as
      a function of rectangular coordinates x and y. Sketch the magnitude of
      $u_z (x, y = 0, z = 0)$ along the x-axis between $x = -4a$ and $x = +4a$.

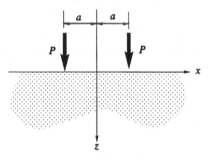

3.4   In Section 3.1, we demonstrated how the stress components found by
      Boussinesq acting on a hemispherical surface must exactly equilibrate
      the point load P. The same conclusion must be true for any surface that
      isolates the point load, such as the cylindrical surface with flat base
      shown below. Find the vertical components of traction that act on the
      sides and base of the cylinder. Integrate these to show that their com-
      bined resultant exactly equilibrates the load P. Show that this conclu-
      sion continues to hold if the diameter of the cylinder approaches in-
      finity, isolating a slab of thickness h. Show that it also holds if the
      cylinder length h approaches infinity.

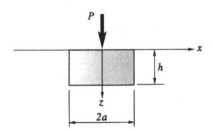

3.5   Use the stress components in eqs. (3.18) and (3.19), together with the line-load geometry shown below, to verify that the traction acting on the cylindrical surface of radius b is given by eq. (3.20).

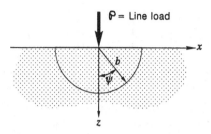

3.6   For the line-load geometry of Question 3.5, demonstrate that P is equilibrated by the vertical resultant of tractions that act on the cylindrical surface of radius b.

3.7   Flamant's problem is an example of plane-strain conditions, where no displacement occurs in one coordinate direction and no quantity depends upon that coordinate. For the plane-strain geometry shown in Figure 3.7, the strain ϵ_{yy} is zero but the stress σ_{yy} is not. Use Hooke's law together with the relations between elastic constants in Table 2.1 to prove that for all plane-strain problems (in the y = constant plane) the stress σ_{yy} will always be given by

$$\sigma_{yy} = \nu(\sigma_{xx} + \sigma_{zz}).$$

3.8   Consider a uniform line load with intensity P acting in the z-direction in an infinite elastic space as shown below. The load lies on the z-axis. Something like this might occur if we had a rigid rod embedded in the

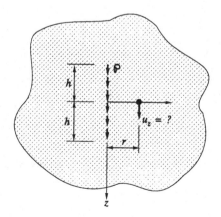

elastic space, and we pulled on the rod. Use Kelvin's solution to show that the z-component of displacement located a distance r from the midpoint of the load (see the figure) is given by

$$u_z(r) = \frac{\wp}{4\pi G(1 - \nu)} \left[\frac{3 - 4\nu}{(1 + r^2/h^2)^{1/2}} + \frac{1}{3(1 + r^2/h^2)^{3/2}} \right].$$

What peculiar thing happens to u_z if we let h approach infinity?

3.9   For Kelvin's problem, show that the resultant upward force acting on a sphere of radius a centered at the origin equilibrates the applied load $2P$.

3.10   In an elastic half-space, the angle θ is defined by $\tan\theta = x/z$, as shown in the sketch below. Demonstrate that the function $\Phi = \theta(x^2 + z^2)$ satisfies the biharmonic equation for all x and z.

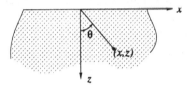

References

Original developments for the point load solutions are found in

Boussinesq, J., "Équilibre d'élasticité d'un solide isotrope sans pesanteur, supportant différents poids," *C. Rendus Acad. Sci. Paris*, Vol. 86, pp. 1260–1263 (1878).

Cerruti, V., "Sulla deformazione di uno strato isotropo indefinito limitato da due piani paralleli," *Atti Acad. Nazl. Lincei. Rend.*, Serie 4, Vol. 1, pp. 521–522 (1884–85).

Flamant, A.A., "Sur la repartition des pressions dans un solide rectangulaire chargé transversalement," *Compte Rendu à l'Académie des Sciences*, Vol. 114, pp. 1465–1468 (1892).

Mindlin, R.D., "Force at a point in the interior of a semi-infinite solid," *Physics*, Vol. 7, pp. 195–202 (1936).

Thompson, W. (Lord Kelvin), "On the equations of equilibrium of an elastic solid," *Cambr. Dubl. Math. J.*, Vol. 3, pp. 87–89 (1848).

Each of the four point-load solutions, Boussinesq's, Kelvin's, Cerrutti's, and Mindlin's, are derived in

Westergaard, H. M., *Theory of Elasticity and Plasticity*, Harvard University Press, Massachusetts (1952).

The other fundamental solutions cited in Section 3.6 are found in

Booker, J. R., Balaam, N. P., and Davis, E. H., "The behavior of an elastic non-ho-
mogeneous half-space," Part 1 — line and point loads, *Int. Jour. Numer. Anal.
Methods Geomech.*, Vol. 9, pp. 353–367 (1985).
Burmister, D. M., "The theory of stresses and displacements in layered systems and
applications to the design of airport runways," *Proc. Highway Res. Board*, Vol.
23, pp. 127–148 (1943).
Lamb, H., "On the propagation of tremors over the surface of an elastic solid," *Phil.
Trans. Roy. Soc. London*, Series A, Vol. 203, pp. 1–42 (1904).
Michell, J. H., "The stress in an aeolotropic elastic solid with an infinite plane
boundary," *Proc. London Math. Soc.*, Vol. 32, pp. 247–258 (1900).
Stokes, G. G., "On the dynamical theory of diffraction," *Trans. Camb. Phil. Soc.*,
Vol. 9, pp. 243–280 (1849).

The Airy stress function was originally given in

Airy, G. B., "On the strains in the interior of beams," *Brit. Assoc. Adv. Sci. Rept.*,
pp. 82–86 (1862).

4

Applications of fundamental solutions

4.1 Introduction

We are now going to see how the fundamental solutions (especially Boussinesq's) can be used to find useful answers for certain geotechnical problems. The main things we will be concerned with are the stresses and displacements caused by a shallow structural foundation. Footings, rafts, grillage foundations, practically all shallow foundations can be treated, provided elastic behavior is a reasonable assumption for the soils involved.

Generally speaking, we can divide foundation problems into two classes. These are *interactive problems* and *noninteractive problems*. In the former category are all problems in which the elasticity of the foundation plays an important role. For example, a flexible raft foundation supporting a multistory structure, like that illustrated in Figure 4.1, *interacts* with the soil. Thinking in terms of what we know about elasticity and structural mechanics, we see the deformation of the raft and the deformation of the soil must both obey requirements of equilibrium and must also be geometrically compatible. If a point on the raft is displaced relative to another point, then we realize bending stresses will develop within the raft and there will be different reactive pressures in the soil beneath those points. The response of the raft and the response of the soil are coupled and must be considered together. In contrast, noninteractive problems are those where we can reasonably assume the elasticity of the foundation itself is unimportant to the overall response of the soil.

Examples of noninteractive problems are illustrated in Figure 4.2. These are situations where the structural foundation is either very flexible or very rigid when compared with the soil elasticity. In noninteractive problems, we don't need to consider the stress-strain response of the foundation. The soil deformations are controlled by the contact pressures exerted by the foundation, but the response of the soil and the structure are effectively uncoupled. It is clear that noninteractive problems will be much simpler than interactive

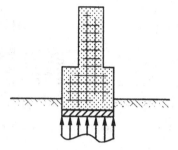

Figure 4.1 A multistory structure supported on a flexible raft foundation: an interactive problem.

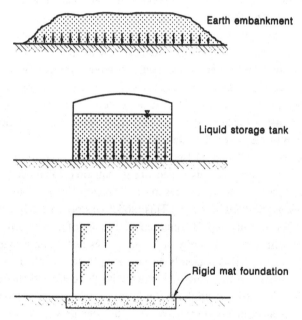

Earth embankment

Liquid storage tank

Rigid mat foundation

Figure 4.2 Some noninteractive problems.

ones. Because of their convenience and simplicity, we will devote most of this chapter to noninteractive problems.

The simplest problems involve a uniform vertical stress applied at the surface of a homogeneous, isotropic, elastic half-space. If the stress acts on an area of regular geometric shape, such as a circle or a square, then the problem is further simplified. Generally, we can fully determine some (if not all)

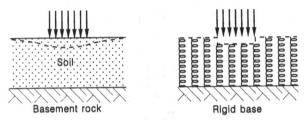

Figure 4.3 Comparison of a Winkler soil model (right) with realistic conditions (left).

of the stresses and displacements just by integrating Boussinesq's fundamental equations over the region covered by the load. The superposition principle allows this integration. It is superposition that makes linear elasticity so powerful in tackling this type of problem.

There are several variations on this simplest case. We can consider nonuniform loads, and these solutions will suggest how to approach the problem of rigid foundations. We can allow the load to extend to infinity on a strip of finite width and have plane-strain conditions. We can consider layered soils, using a simple method of analysis that is not exact but provides good approximations for displacements. And we can make some generalizations about consolidation of clays and time-dependent settlements.

There are other methods for finding appropriate solutions to these types of foundation problems. They use specialized mathematical models for soils, models that mimic some of the response characteristics of linear elasticity. The simplest is the *Winkler model*. This model was proposed by the German engineer E. Winkler in 1867. He represented the case of a finite soil layer resting on basement rock by a family of linear springs resting on a rigid base, as shown in Figure 4.3. This is clearly an over-simplified model. If we compare the deflected shape of the real soil surface with the deflected shape of Winkler's model we find some similarity, but there are significant differences. The Winkler springs act independently. We can push one down, and its neighbors are not affected. This is unlike the real soil. The deflected shape in the real soil, or in a linear elastic model for the real soil, is continuous. Despite this fundamental difference, Winkler models are extremely useful in many types of soil-foundation interaction analyses.

A Winkler model requires only one parameter, the elastic spring constant. A number of researchers have modified the original model in an effort to obtain a more realistic deflected shape. To do this, an additional parameter is needed, and these modified models are usually referred to as two-parameter elastic models. One simple two-parameter model, which also has an intuitive physical ex-

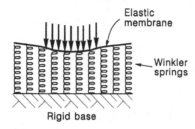

Figure 4.4 The two-parameter model of Filonenko-Borodich.

planation, is shown in Figure 4.4. This is the model of Filonenko-Borodich. An elastic membrane is stretched across the surface of the Winkler springs. The membrane partially connects the springs depending on how tightly it is initially stretched. The two parameters here are the spring stiffnesses and the membrane tension. There are several other two-parameter models. They are discussed in detail, as is the Winkler model, in Selvadurai's *Elastic Analysis of Soil-Foundation Interaction*, which we referred to in Chapter 1.

The plan for this chapter will be to consider some simple problems involving a homogeneous elastic half-space, then to add a little complexity by way of nonuniform loadings and layered soils, and also to consider consolidation and time-dependent settlements. Finally, we will look at some specialized elastic analyses that are useful in in situ testing and in determination of gravity stresses. We will employ compression positive notation throughout this chapter.

4.2 Uniform circular load on homogeneous half-space

This is the simplest possible application. It might arise in analysis of a liquid storage tank like that shown in Figure 4.2(b). We can represent the problem as a vertical pressure p_o applied uniformly over the circular region $\mathcal{R}$ of radius a as shown in Figure 4.5. The displacement or the stress at any point in the half-space can be determined by integrating Boussinesq's solution over the loaded region. We will begin by considering the displacement of the ground surface $z = 0$.

Referring back to eqs. (3.4), we see that the displacements on the surface $z = 0$ for the vertical point load are

$$u_r = -\frac{P(1 - 2\nu)}{4\pi Gr}, \quad u_\theta = 0, \quad u_z = \frac{2P(1 - \nu)}{4\pi Gr}. \tag{4.1}$$

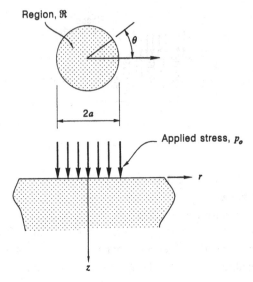

Figure 4.5 Uniform stress p_o applied over a circular region on the half-space surface.

Note that when $z = 0$, R and r become equal. What we want to do now is to replace the point load P by the applied stress p_o multiplied by an element of area dA. This "point load," $p_o\,dA$, will contribute to the displacements at all points on the half-space surface. If we select one particular point, then we can add together all of these contributions by integrating over the whole region $\mathfrak{R}$. To illustrate this, consider the point exactly under the center of the tank. The contribution to the vertical displacement u_z at this point will be

$$du_z = \frac{(p_o rd\theta dr)(1 - \nu)}{2\pi Gr}. \tag{4.2}$$

Here $rd\theta dr$ is the element of area. The product $p_o rd\theta dr$ plays the role of the point load P in (4.1). If we integrate both sides of this equation we find the total vertical displacement, or settlement, at the center of the tank

$$u_z(r = 0, z = 0) = \iint\limits_{\mathfrak{R}} \frac{(p_o rd\theta dr)(1 - \nu)}{2\pi Gr}$$

$$= \int_0^{2\pi} \int_0^a \frac{p_o(1 - \nu)}{2\pi G}\,drd\theta \tag{4.3}$$

$$= \frac{p_o a(1 - \nu)}{G}.$$

This simple result gives us the surface settlement at the center of any uniform circular load.

In calculating u_z in (4.3) we have side-stepped an interesting question. In Boussinesq's solution, the point load lies at the origin of coordinates. In our circular load problem, we've held the coordinate origin at the center of the loaded region while we *moved* the point load over the region of integration. Is this important? The answer is no, if we are dealing with the vertical displacement u_z, but would be yes if we were looking for the radial displacement u_r. In Boussinesq's solution the point load acts vertically. The contribution du_z due to the "point load" $p_o rd\theta dr$ is always in the vertical direction. If we were dealing with radial displacements, however, the contribution du_r would always point toward the place where $p_o rd\theta dr$ happened to lie. As we move the "point load" over the region $\mathcal{R}$, the contribution keeps changing in direction. So we see that the vertical displacement is simpler to deal with in these circumstances, which is convenient, of course, because the vertical settlement is usually the most interesting quantity in foundation analysis.

Next, consider the vertical displacement at the edge of the circular load. We will move the origin of coordinates to a point on the circumference of the circle as shown in Figure 4.6. Now the radial distance between the point load and the origin is s. The element of area is $sd\Omega ds$. The contribution to the vertical displacement will be

$$du_z = \frac{(p_o sd\Omega ds)(1 - \nu)}{2\pi Gs}.$$

Integrating both sides of this equation we have

$$u_z = \iint_{\mathcal{R}} \frac{(p_o sd\Omega ds)(1 - \nu)}{2\pi Gs}.$$

The limits of integration for Ω are $-\pi/2$ and $\pi/2$, while s varies between 0 and $2a \cos\Omega$

$$u_z = \int_{-\pi/2}^{\pi/2} \int_0^{2a\cos\Omega} \frac{p_o(1 - \nu)}{2\pi G} \, dsd\Omega$$

$$= \int_{-\pi/2}^{\pi/2} \frac{p_o(1 - \nu)}{2\pi G} (2a \cos\Omega)d\Omega$$

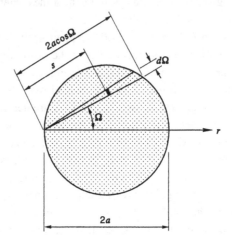

Figure 4.6 Geometry for finding displacement at edge of loaded region.

so that

$$u_z(r = a, z = 0) = \frac{2p_o a(1 - \nu)}{\pi G}.$$ (4.4)

This is less than the displacement at the center by a factor of $2/\pi$.

Since we are so often interested in the surface displacements it is convenient to introduce a special symbol to use in place of $u_z(z = 0)$. We will let w play this role. It will always denote the displacement or settlement of the free surface. In axisymmetric problems such as the circular load, w is only a function of the radial position, $w = w(r)$. In other problems, w will generally depend upon two coordinates, $w = w(x,y)$. Restating the results just obtained

$$w(r = 0) = \frac{p_o a(1 - \nu)}{G}, \quad w(r = a) = \frac{2p_o a(1 - \nu)}{\pi G}.$$ (4.5)

Now what about other points besides $r = 0$ and $r = a$? For an arbitrary point inside the circle, we can use the geometry set out in Figure 4.7. Studying this geometry we find

$$L_1 = a(\sqrt{1 - \alpha^2\sin^2\Omega} - \alpha\cos\Omega),$$

$$L_2 = a(\sqrt{1 - \alpha^2\sin^2\Omega} + \alpha\cos\Omega).$$

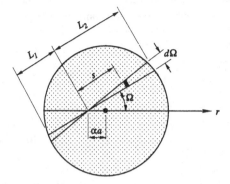

Figure 4.7 Geometry for finding displacement at interior point of loaded region.

Here α plays the role of dimensionless distance, r/a, from the center of the circle. The integration now looks like this

$$w(\alpha a) = \iint_{\mathscr{R}} \frac{(p_o s d\Omega ds)(1 - \nu)}{2\pi G s}$$

$$= \int_{-\pi/2}^{\pi/2} \int_0^{L_1} \frac{p_0(1 - \nu)}{2\pi G} \, ds d\Omega + \int_{-\pi/2}^{\pi/2} \int_0^{L_2} \frac{p_0(1 - \nu)}{2\pi G} \, ds d\Omega.$$

Because of symmetry about the diameter of the circle, we need only to integrate from 0 to $\pi/2$ for Ω and then double the answer. We can write

$$w(\alpha a) = \frac{p_0(1 - \nu)}{\pi G} \left[\int_0^{\pi/2} L_1 d\Omega + \int_0^{\pi/2} L_2 d\Omega \right]$$

which simplifies to

$$w(\alpha a) = \frac{2p_0(1 - \nu)a}{\pi G} \int_0^{\pi/2} \sqrt{1 - \alpha^2 \sin^2\Omega} \, d\Omega, \quad \text{for } \alpha \le 1. \quad (4.6)$$

The integral here is called a complete *elliptic integral*. It can not be integrated in closed form, but, because they frequently occur in many physical problems, elliptic integrals have been tabulated for values of the parameter α. Tables of elliptic integrals are found in many mathematical handbooks. Two particular values of α are simple, however. Those are $\alpha = 0$ and $\alpha = 1$, and if we substitute those values we find the settlements at the center and at the edge, exactly the same as in (4.5).

For points outside the circle (with $\alpha > 1$) a similar analysis can be used. More elliptic integrals appear. The result looks like this

$$w(\alpha a) = \frac{2p_o(1 - \nu)a}{\pi G} \left\{ \alpha \int_0^{\pi/2} \sqrt{1 - \frac{1}{\alpha^2} \sin^2\Omega} \; d\Omega \right.$$

$$\left. -\left(\alpha - \frac{1}{\alpha}\right) \int_0^{\pi/2} \frac{d\Omega}{\sqrt{1 - \frac{1}{\alpha^2} \sin^2\Omega}} \right\}, \quad \text{for } \alpha \geq 1. \tag{4.7}$$

We can introduce a dimensionless settlement $W = W(\alpha)$ defined by

$$W(\alpha) = \frac{w(\alpha a)G}{p_o(1 - \nu)a}. \tag{4.8}$$

A graph of W for values of α between 0 and 3 is shown in Figure 4.8. By removing dimensions, W effectively gives the settlement for any circle size and any applied pressure and for any combination of elastic constants. We see that W is greatest beneath the center of the load and diminishes rapidly outside the loaded circle. Sometimes this shape is referred to as a *settlement bowl*.

Next, consider the vertical displacement on the z-axis beneath the center of the circular load. Recall the Boussinesq solution for u_z

$$u_z(r,z) = \frac{P}{4\pi GR} \left[2(1 - \nu) + \frac{z^2}{R^2} \right]. \tag{4.9}$$

To use this expression we can replace P by $p_o r d\theta \, dr$ and integrate over the circular region. Referring to Figure 4.9, the contribution to u_z at depth z beneath the center will be

$$du_z = \frac{(p_o r d\theta dr)}{4\pi GR} \left[2(1 - \nu) + \frac{z^2}{R^2} \right]$$

and integrating

$$u_z = \int_0^{2\pi} \int_0^a \frac{p_o}{4\pi G} \left[2(1 - \nu) \frac{r}{R} + \frac{z^2 r}{R^3} \right] dr d\theta.$$

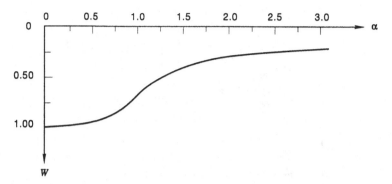

Figure 4.8 Dimensionless settlement profile for the uniform circular load; the *settlement bowl.*

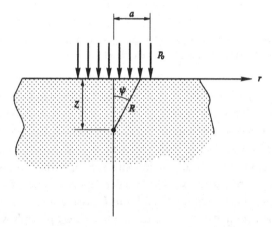

Figure 4.9 Geometry for finding displacement of a point at depth z beneath the center of the uniform circular load.

We can carry out the θ-integration immediately. To help with the r-integration, we introduce the angle ψ shown in Figure 4.9. The integral becomes

$$u_z = \frac{p_o}{2G} \int_0^{\psi_0} [2(1 - \nu)\,\sin\psi - \sin\psi\,\cos^2\psi]z\,\sec^2\psi\,d\psi$$

where $\psi_o = \tan^{-1}(a/z)$. Integrating this we find the vertical displacement at depth z

$$u_z(z) = \frac{p_o}{2G}\left\{2(1 - \nu)(\sqrt{a^2 + z^2} - z) - z\left(\frac{z}{\sqrt{a^2 + z^2}} - 1\right)\right\}. \quad (4.10)$$

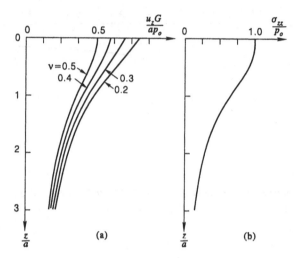

Figure 4.10 Variations of (a) dimensionless displacement, and (b) dimensionless stress with depth beneath center of uniform circular load.

A graph of this equation in dimensionless form is shown in Figure 4.10(a) for several values of Poisson's ratio.

Now we shift our attention to the stress field in the half-space. We can use the same technique, replacing P with $p_o r d\theta dr$ and integrating over $\Re$, to find the components of stress. The simplest stress component and also the most interesting, is σ_{zz}, the vertical stress. Consider the distribution of σ_{zz} along the z-axis, beneath the center of the load. From Boussinesq's solution, eq. (3.5), we see that the contribution to σ_{zz} from the *point load* $p_o r d\theta dr$ will be

$$d\sigma_{zz} = \frac{3}{2\pi} (p_o r d\theta dr) \frac{z^3}{R^5}.$$

We can integrate this using the same scheme we used for the vertical displacement u_z (as in Figure 4.9)

$$\sigma_{zz} = \frac{3p_o}{2\pi} \int_0^{2\pi} \int_0^a \frac{rz^3}{R^5} dr d\theta$$

$$= 3p_o \int_0^{\psi_0} \sin\psi \cos^2\psi \, d\psi \qquad (4.11)$$

$$= p_o \left[1 - \frac{z^3}{(a^2 + z^2)^{3/2}} \right].$$

A graph of this equation, in dimensionless form, is shown in Figure 4.10(b). We see from this figure, and also from eq. (4.11), that σ_{zz} approaches p_o as z approaches zero, just as we would expect.

Values of σ_{zz} at other points than those on the z-axis may be found by using similar methods to those described for u_z, and, as one would expect, elliptic integrals again appear. Other components of stress can also be determined, but just as was the case with the radial displacement u_r, we must be careful with the orientation between the "point load" in the integration and the directions in which the stress components act. This is true of all the remaining nonzero stresses: σ_{rr}, $\sigma_{\theta\theta}$, and σ_{rz}, except on the z-axis where they all vanish. In contrast to individual stress components, the stress invariants discussed in Section 1.8 do not depend on our choice of coordinate system. They can be found relatively easily by integration over the loaded region.

4.3 Uniform loads of other shapes, homogeneous half-space

The uniform circular load just considered can be looked on as a prototype problem that we will vary in different ways. One important variation is to consider loaded regions of other shapes. We will stay with our homogeneous half-space for the time being and be primarily concerned with the vertical surface displacement w and the vertical stress σ_{zz}.

Consider the triangular-shaped load with constant magnitude p_o illustrated in Figure 4.11. This is a right triangle (later we can drop that restriction) and two sides are aligned with the x- and y-axes. Figure 4.11 shows the plan of the load, looking down on the surface of the half-space. We will first find the settlement w beneath the corner of the triangle marked A. The "point load" here will be $p_o dx dy$ and the contribution to w will be

$$dw = \frac{(p_o dx\, dy)(1 - \nu)}{2\pi G\sqrt{x^2 + y^2}} \tag{4.12}$$

which follows from eq. (4.1). We integrate both sides of this equation to find the settlement at point A

$$w = \int_0^a \int_0^{\left(\frac{b}{a}\right)x} \frac{p_o(1 - \nu)}{2\pi G\sqrt{x^2 + y^2}}\, dy\, dx$$

$$= \int_0^a \frac{p_o(1 - \nu)}{2\pi G} \left[\ell_n(y + \sqrt{x^2 + y^2}) \right]_{y=0}^{y=\left(\frac{b}{a}\right)x} dx.$$

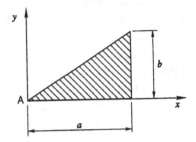

Figure 4.11 Triangular loaded region.

Now we see, when we evaluate the y-integral, that x cancels from the integrand altogether, making the second integration trivial. We find

$$w = \frac{p_o(1 - \nu)a}{2\pi G} \, \ell_n\!\left(\frac{b}{a} + \sqrt{1 + \frac{b^2}{a^2}}\right). \qquad (4.13)$$

We can write this in another, equivalent, form

$$w = \frac{p_o(1 - \nu)a}{2\pi G} \, \sinh^{-1}\!\left(\frac{b}{a}\right). \qquad (4.14)$$

At this point, you might well wonder why we want to consider a loaded region in the shape of a right triangle at all. There may not be many real-life applications for loads of this shape. The answer to this objection is that we can use the triangular region so effectively to build up, using superposition, a great range of other shapes. For example, to find the settlement beneath one corner of a rectangular load, we just need to put together two triangles, as in Figure 4.12. The settlement at A will consist of two parts

$$w = w_1 + w_2$$

where w_1 in the settlement is due to triangle 1

$$w_1 = \frac{p_o(1 - \nu)a}{2\pi G} \, \sinh^{-1}\!\left(\frac{b}{a}\right)$$

and w_2 the settlement due to triangle 2

$$w_2 = \frac{p_o(1 - \nu)b}{2\pi G} \, \sinh^{-1}\!\left(\frac{a}{b}\right).$$

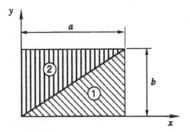

Figure 4.12 Superposition of two triangular loads to create a rectangular loaded region.

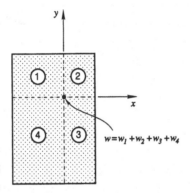

Figure 4.13 Superposition of four small rectangles to find the settlement at an interior point of a rectangular load.

We see that w_1 and w_2 have the roles of a and b reversed to account for the position of the corner A relative to the two sides of the appropriate triangle. We put the two parts together to have the settlement due to the rectangular load

$$ w = \frac{p_o(1 - \nu)a}{2\pi G} \left[\sinh^{-1}\left(\frac{b}{a}\right) + \frac{b}{a} \sinh^{-1}\left(\frac{a}{b}\right) \right]. \qquad (4.15) $$

For the special case of a square, set $a = b$, then

$$ w = \frac{p_o(1 - \nu)a}{\pi G} \sinh^{-1}(1) = 0.281 \frac{p_o(1 - \nu)a}{G}. \qquad (4.16) $$

So far we have considered vertical displacements beneath one corner of a loaded region, either a triangle or a rectangle. To find the settlement at any other point we can use superposition. For example, the settlement at an interior point of the rectangular load is given by the sum of the settlements at the

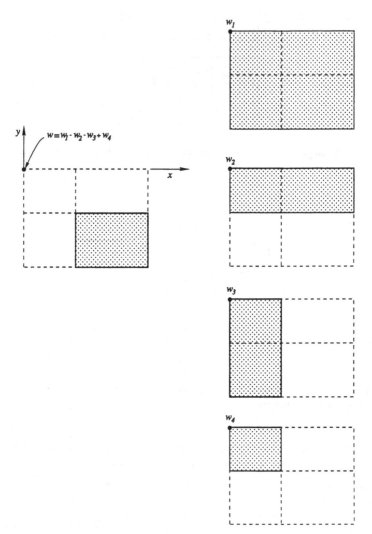

Figure 4.14 Using superposition to find the settlement at a point exterior to a rectangular load.

corners of four subrectangles, as shown in Figure 4.13. To find the settlement at a point outside the loaded region, we can add and subtract corner settlements for four rectangles as shown in Figure 4.14. We first find the settlement w_1 as if the entire region defined by the actual load and the point of interest were covered by the applied stress. We then subtract the settlements due to the two edge rectangles, w_2 and w_3. Finally, we have to add back in the settlement w_4 due to the smallest rectangle since it has been subtracted twice in w_2 and w_3.

Our solution for the right triangle, eqs. (4.13) and (4.14), can also be generalized to any triangular-shaped load using superposition. Figure 4.15 shows how the value of w beneath one corner of any triangular load is found from the displacements associated with two right triangles. Figure 4.16 shows how these solutions can be further generalized to find the displacement at any point by superposing corner displacements for a number of triangles. We can now easily see how the surface displacement at any point due to a uniform load applied over any polygonal-shaped area can be built up from a sequence of solutions for triangular loads.

Next, turn attention to the stress field. The most interesting component of stress is σ_{zz}, and, as we discussed before, it is also the simplest to deal with. Consider the value of σ_{zz} on the vertical line below one corner of a uniform load acting on a right triangular region. We can use the geometry shown in Figure 4.11. From Boussinesq's solution, eq. (3.5), the contribution to σ_{zz} due to the "point load" $p_o dx dy$ will be

$$d\sigma_{zz} = \frac{3(p_o dx dy)z^3}{2\pi[x^2 + y^2 + z^2]^{5/2}}$$

which after integration yields

$$\sigma_{zz}(z) = \frac{p_o}{2\pi}\left[\tan^{-1}\left(\frac{ca}{bz}\right) - \tan^{-1}\left(\frac{a}{b}\right) + \frac{abz}{(a^2 + z^2)c}\right] \qquad (4.17)$$

where

$$c = \sqrt{a^2 + b^2 + z^2}.$$

We can now superpose triangular-load solutions to find σ_{zz} at any depth z due to any polygonal-shaped uniform load. For example, the stress beneath one corner of a rectangular load is found by superposing two triangular loads as in Figure 4.12. We find

$$\sigma_{zz} = \frac{p_o}{2\pi}\left[\tan^{-1}\left(\frac{cb}{az}\right) + \tan^{-1}\left(\frac{ca}{bz}\right) - \tan^{-1}\left(\frac{b}{a}\right)\right.$$
$$\left. - \tan^{-1}\left(\frac{a}{b}\right) + \frac{abz}{c}\left(\frac{1}{\ell^2} + \frac{1}{m^2}\right)\right] \qquad (4.18)$$
$$= \frac{p_o}{2\pi}\left[\tan^{-1}\left(\frac{ab}{cz}\right) + \frac{abz}{c}\left(\frac{1}{\ell^2} + \frac{1}{m^2}\right)\right]$$

where $\ell^2 = a^2 + z^2$ and $m^2 = b^2 + z^2$.

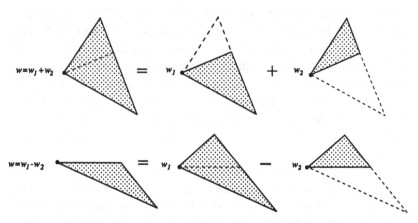

Figure 4.15 Using superposition to find the settlement at one corner of a non-right-triangle load.

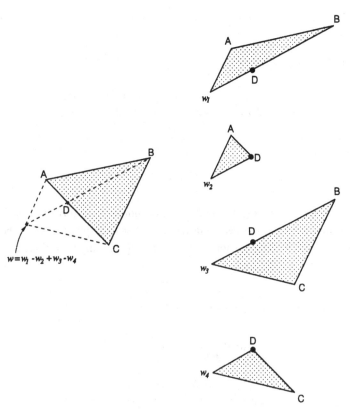

Figure 4.16 Using superposition to find the settlement of a point exterior to a triangular load.

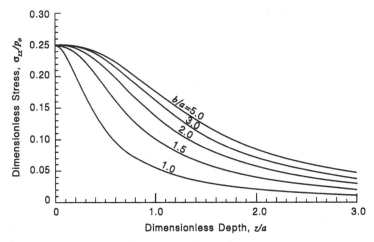

Figure 4.17 Dimensionless stress beneath one corner of a rectangular loaded area that has sides of length a and b.

Equation (4.18) can be used with superposition to find the vertical stress at any point in an elastic half-space due to any uniform rectangular load, exactly as the vertical displacement w was determined in Figures 4.13 and 4.14. Equation (4.18) can also be represented graphically, in dimensionless form, as shown in Figure 4.17. Graphs similar to Figure 4.17 can be found in most textbooks dealing with foundation engineering.

4.4 Non-uniform loads, homogeneous half-space

Next we need to consider applied stresses that are functions of position. We replace p_o by $p(x,y)$, a function of x and y, over some region $\mathcal{R}$ of the half-space surface. The only difference between this problem and those considered in the previous two sections is that the elemental "point load" $p(x,y)dxdy$ is now a function of position, which may possibly complicate the integrations.

As an example, consider the parabolic stress distribution applied over a circular region, illustrated in Figure 4.18. The applied stress is given by

$$p(r) = p_o\left(1 - \frac{r^2}{a^2}\right) \quad \text{for } 0 \le r \le a. \quad (4.19)$$

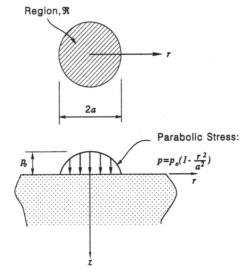

Figure 4.18 Parabolic stress distribution over a circular region.

If we wished to determine the surface displacement beneath the center of the load, we would proceed as in eq. (4.3), but now we have

$$w(r = 0) = \iint_{\mathcal{R}} \frac{(p(r)rdrd\theta)(1 - \nu)}{2\pi Gr}$$

$$= \int_0^{2\pi} \int_0^a \frac{p_o(1 - r^2/a^2)(1 - \nu)}{2\pi G} \, drd\theta \qquad (4.20)$$

$$= \frac{2}{3} \frac{p_o a(1 - \nu)}{G}.$$

A second example is the inverted parabolic stress distribution shown in Figure 4.19. The applied stress is specified by

$$p(r) = p_o + \beta p_o \frac{r^2}{a^2} \quad \text{for } 0 \le r \le a \qquad (4.21)$$

and the surface settlement beneath the center of the loaded area is easily found to be

$$\dot{w}(r = 0) = \frac{p_o a(1 - \nu)}{G} \left(1 + \frac{\beta}{3}\right). \qquad (4.22)$$

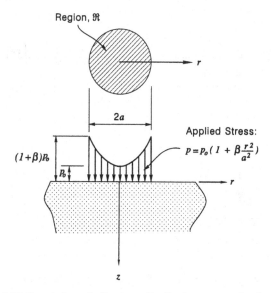

Figure 4.19 Inverted parabolic stress distribution over a circular region.

These two examples are of particular interest because experiments have revealed stress distributions beneath rigid circular foundations that are similar to both. Experiments involving sands show the parabolic distribution, while experiments with clay soils show the inverted parabolic shape. The reasons that underlie these findings are explored in the next section, once we have discussed rigid foundations a little more fully.

4.5 Rigid foundations

So far in this chapter we have considered problems where the stress distribution on the surface of an elastic half-space has been fully specified in advance. Physically, these problems correspond to structural foundations that are flexible relative to the stiffness of the underlying soils. For example, the structure might be an embankment, such as is illustrated in Figure 4.20(a). We would be justified in assuming a stress distribution that would equal the gravity stresses exerted by the embankment on the underlying soil. In contrast, consider a thick concrete footing supporting a column load $\mathcal{P}$ like that illustrated in Figure 4.20(b). Here, the stress distribution is unknown, except for the constraint that the resultant of the contact stresses must equal $\mathcal{P}$. The thing we do know in this problem is that the footing (if it is sufficiently thick) will be far stiffer than the soil and the displacements beneath the footing will be more or less the same at all points. Here it is reasonable to assume the footing is perfectly rigid.

Applications of fundamental solutions

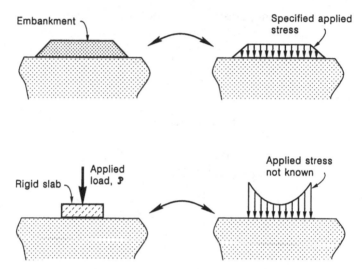

Figure 4.20 Contrasting flexible and rigid foundations.

Rigid foundation problems are more difficult to deal with than the problems we have encountered so far. The contact stresses have a nasty habit of becoming singular at some points, and the displacements may not change smoothly as we move from point to point on the half-space surface. We can illustrate all the relevant features of the rigid foundation problem by considering a rigid circular footing like that shown in Figure 4.21. We will assume the base of the footing is frictionless so the soil is free to move horizontally, and we require the vertical displacement to be equal to δ at all points beneath the footing.

One way to look at the problem is as follows: We ask ourselves, what stress distribution $p(r)$ will give a constant displacement $w = \delta$ at all points in the circular region $0 \leq r \leq a$? If we could find a stress distribution that would accomplish this, and if, moreover, its resultant were equal to $\mathcal{P}$

$$\int_0^{2\pi} \int_0^a p(r)rdrd\theta = \mathcal{P} \tag{4.23}$$

then we would have a solution to the entire problem. The vertical displacement beneath the footing would necessarily equal δ, and we could find the displacements and stresses at any point in the half-space by integrating the appropriate equation, just as was done earlier in this chapter. The trick, of course, is to find the right stress distribution. Boussinesq found it first, and in a moment we will write down his solution; but first, consider what guesses one might make about $p(r)$, knowing only what has been pointed out in this paragraph. We could safely guess that the stress must be higher near the edge

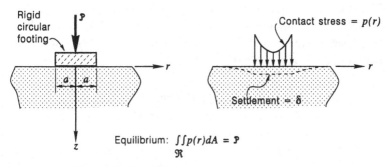

Figure 4.21 The problem of a rigid circular footing.

of the footing than at the center, because we want the vertical displacement to be the same everywhere beneath the footing, and we already know that if the stress were uniform the displacement at the center would be about 50% greater than at the edge. Thinking a little deeper, we might anticipate the stress at the edge could be infinite. This is suggested by the *kink* in the surface displacement profile that we expect to find at the edge of the footing. The surface displacements (we would expect) will increase smoothly as we move toward the footing from some point outside. The slope of the surface becomes steeper and steeper, but, at the footing edge, everything changes abruptly. The displacements become constant, and the slope has a discontinuity. From all the solutions for the stress distributions discussed earlier, it is hard to see how we could make this kink occur. Our earlier solutions have all had smoothly changing displacements. The one thing we have not investigated earlier is the result of the stress being infinite, and that might suggest guessing a stress singularity at the point where the kink occurs, the footing edge. So if we were to guess $p(r)$, we might look at functions having a minimum at $r = 0$ and increasing to infinity as r approaches a. Moreover, we also have to satisfy eq. (4.23). It tells us that even though the stress may be singular, its resultant must be finite and equal to $\mathcal{P}$.

As we noted above, Boussinesq first solved this problem. The correct stress distribution is given by

$$p(r) = \frac{\mathcal{P}}{2\pi a} \frac{1}{\sqrt{a^2 - r^2}} \quad \text{for } 0 \le r \le a. \tag{4.24}$$

We see here what we expected to find. The stress is a minimum at $r = 0$ and increases to infinity as r approaches a. We can also verify that (4.23) is satisfied by carrying out the necessary integrations.

The stress distribution given in (4.24) will result in a constant settlement $w = \delta$ beneath the footing. To see how δ is related to $\mathcal{P}$, we can use Boussinesq's solution and integrate the given stress distribution. The easiest point at which to find δ is at the center:

$$w(r = 0) = \delta = \iint_{\mathcal{R}} \frac{(p(r)rdrd\theta)(1 - \nu)}{2\pi Gr}$$

$$= \int_0^{2\pi} \int_0^a \frac{\mathcal{P}(1 - \nu)}{4\pi^2 Ga} \frac{drd\theta}{\sqrt{a^2 - r^2}} \qquad (4.25)$$

$$= \frac{\mathcal{P}(1 - \nu)}{4Ga}.$$

We see that δ depends only on $\mathcal{P}$, the footing radius a, and the elastic properties of the half-space. If we were to calculate w at any other point beneath the footing, the integrations would be more complicated, but the result $w = \delta$ would be the same. For points outside the circular footing, the settlement dies out slowly

$$w(r) = \frac{\mathcal{P}(1 - \nu)}{2\pi Ga} \sin^{-1}\left(\frac{a}{r}\right) \quad \text{for } a \leq r.$$

Clearly the rigid-foundation problem is more difficult to deal with than were the flexible foundation, uniform stress, problems we considered earlier. It is natural to wonder just how different the surface settlements are for similar total loads on rigid and flexible footings. Results for circular footings are shown in Figure 4.22. The uniform stress p_o is taken to be

$$p_o = \frac{\mathcal{P}}{\pi a^2}$$

so that both cases consider the same total load. Beneath the center of the footing, the uniform stress distribution naturally gives the greatest settlement, about 30% greater than the rigid footing. At the footing edge, the uniform stress settlement is slightly smaller than the rigid footing settlement. The difference is about 20%. For points outside the footing ($r > a$), the two cases grow closer and closer together. For values of r greater than $1.5a$, the difference in surface settlement is less than 5%. We conclude that for certain problems, the exact nature of the load distribution may not be of great importance.

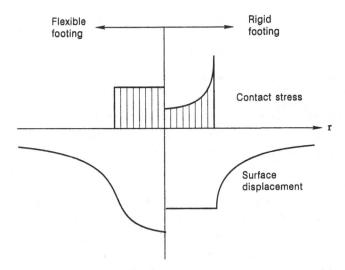

Figure 4.22 Profiles of contact stress and settlement for flexible and rigid circular footings.

These would be problems where the effect of a loaded area on neighboring regions is of more interest than the settlement beneath the load itself. If points of special interest lie well away from the loaded area, then the stress distribution can probably be assumed to be uniform without any appreciable error. This theme is developed much more fully in Appendix E.

Rigid square and rectangular foundations are unfortunately not so amenable to analysis as are circular footings. Many investigators have considered rigid rectangular footings using a variety of approximate methods. The contact stresses become singular at the footing edges, exactly as we would expect, but the settlement profile quickly approaches that for a uniform applied stress at points outside the footing. Compare the settlements for a rigid rectangular footing with the settlements at the center and at one corner of a uniformly loaded rectangular area shown in Table 4.1. The average applied stress p_o is the same in both cases, and the settlements are all in this dimensionless form

$$W = \frac{wG}{ap_o(1 - \nu)}.$$

We see from the table that the center and corner settlements for the uniform load bracket the settlements for the rigid footing in all cases. The average of the center and corner values is close to the rigid footing settlements, varying by as little as 3% for $b/a = 1$ up to 7% for $b/a = 5$.

Table 4.1 *Dimensionless settlement of rigid rectangular footings*

b/a =	1.0	1.5	2.0	3.0	5.0
Rigid footing[1]	.4439	.5268	.5984	.7061	.8510
Flexible footing center	.5611	.6788	.7659	.8915	1.0523
Flexible footing corner	.2806	.3394	.3829	.4458	.5262
Flexible footing average	.4209	.5091	.5744	.6687	.7893

[1]Values taken from the paper by Dempsey and Li (1989).

Return now to the question that we foreshadowed at the end of the preceding section. Just how close to reality are these rigid foundation solutions? We have observed that the contact stress becomes infinite at the foundation edges, and we know this cannot occur in a real soil. Moreover, we emphasized earlier that elasticity solutions were useful *so long as* the stress levels were not too great. Are these rigid foundation solutions of any value in soil engineering? The answer is yes, but we need to treat them with care.

First note that at the footing edge our infinite contact stress will cause infinite shearing stress. Both the radial stress σ_{rr} and the azimuthal stress $\sigma_{\theta\theta}$ will remain finite (for an incompressible elastic body they will be zero), and a glance at Mohr's circle shows that all surfaces other than principal surfaces will have infinite shear stress at the footing edge. While we can contemplate infinite stresses in the theory of elastic behavior, we are aware that any real material will possess a finite shear strength that must limit the stress magnitude in the annular region near the edge. We might therefore expect to find a truncated stress distribution more like that indicated in Figure 4.23. The details of this truncation obviously depend on the details of the shear strength of the material involved.

This raises a new question: What do we know about the shear strength of soils that may affect the contact stress distribution? The answer, of course, depends on whether we are dealing with cohesive or cohesionless soil. Cohesionless soils derive their shear strength from interparticle friction and confinement of the granular assembly, which in turn depends directly on the normal stress acting on the surface in question. Now we can begin to see how the parabolic stress distribution illustrated in Figure 4.18 might arise. Consider a rigid circular plate placed on the surface of a deposit of uniform sand. If a load $\mathscr{P}$ is gradually applied to the plate, our elastic analysis suggests the immediate development of a stress singularity. But the sand particles near the footing edge are unconfined or at best poorly confined and therefore possess little shear

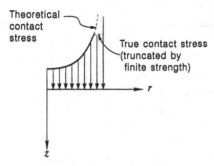

Figure 4.23 Effect of finite strength of soil on theoretical contact stress for rigid circular footing.

strength. The stress singularity must be heavily truncated. Moving from the footing edge toward the center, the degree of confinement increases, shear strength increases, and the amount of truncation due to strength limitations will diminish. Near the center, the true stress will necessarily be greater than the theoretical stress simply in order to equilibrate the applied load $\mathcal{P}$. The resulting stress distribution will look roughly parabolic (as in Figure 4.18), and this is exactly what is observed in experiments on sand.

Now consider the same situation but with a cohesive clay soil rather than cohesionless sand. Shear strength of the clay will be at least partially independent of confining stress, and truncation of the stress singularity at the footing edge will not be nearly so dramatic as in the case of the sand. For a clay with high cohesion, a significant stress increase near the footing edge is possible, and a stress distribution like that illustrated in Figure 4.19 results. Experimental observations support this conclusion.

Considering all that has been said here, one might conclude that elastic analysis of rigid foundations is a waste of time whenever sands are involved. Not so. First, it is through the elastic analysis that we have been able to comprehend the underlying mechanics of a rigid footing on either a cohesionless or cohesive soil. Second, most practical problems involving rigid foundations find the footings buried at some depth beneath the ground surface. The effect of this burial is to increase the average confining stress that acts near the footing, and this implies an increased shear strength, especially for sands. For a buried footing, the contact stress distribution, even in sandy soils, may be expected to look more like the inverted parabolic rather than the parabolic case.

To end this section, we can briefly comment on interactive foundation problems, where finite foundation stiffness plays a role. These problems lie somewhere between perfectly flexible foundations, where we need only specify the applied stresses, and perfectly rigid foundations, which were just discussed. Many investigators have studied interactive problems using a variety of techniques. By far the most popular approach is to use a Winkler foundation, but there are also numerous important developments that analyze flexible foundations on an elastic half-space. These topics are beyond the scope of an undergraduate text such as this, but if you are interested in reading and learning more, see *Elastic Analysis of Soil-Foundation Interaction* by Selvadurai cited in Chapter 1 and in the Rankine Lecture by R.E. Gibson (1974).

4.6 Plane-strain problems

Consider a long narrow rectangular footing such as might support a bearing wall of a structure. We could safely guess that, if the length were considerably greater than the width, the stresses and displacements near the center of the footing will be nearly independent of the footing length. We might then assume the footing is infinitely long, because this will simplify things quite a lot. The resulting conditions, in the context of the theory of elasticity, are called *plane-strain* conditions. To be more precise, suppose the footing was infinitely long, with constant dimensions and constant applied load $\mathcal{P}$ at all points, like the situation illustrated in Figure 4.24. Under these conditions, all the stresses and strains are independent of y, the direction parallel to the footing, and every plane surface defined by $y = $ constant will behave exactly like every other similar surface. To deal with the problem we need to consider only the x-z plane. We mentioned plane-strain problems briefly in Chapter 3 when we analyzed Flamant's line-load problem and when we discussed gravity stresses in Section 3.7. Clearly Flamant's solution will be useful in dealing with the infinitely long footing we have here.

The simplest plane-strain problem is the *strip-load* problem illustrated in Figure 4.25. We have a uniform applied stress p_o acting on an infinitely long strip on the surface of an elastic half-space. We can attack the problem using Flamant's solution, and we will begin by examining the stress field in the half-space. We will postpone considering settlements because they have an unfortunate aspect peculiar to plane-strain problems.

Returning to Figure 4.25, we can consider the strip load as a sequence of line loads standing side by side, each line load having intensity $p_o dx$. Flamant's solution gives us the stress components at any point in the half-space due to the line load. The contribution to the vertical stress $d\sigma_{zz}$ due to the line load is given by eq. (3.18)

$$d\sigma_{zz} = \frac{2(p_o dx)z^3}{\pi(x^2 + z^2)^2}. \qquad (4.26)$$

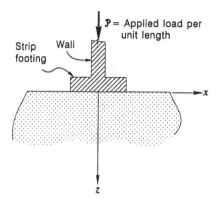

Figure 4.24 Plane-strain conditions: a long strip footing.

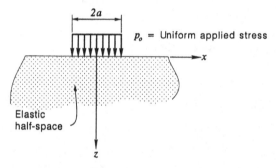

Figure 4.25 Plane-strain conditions: a uniform strip load.

Figure 4.26 illustrates the line load and the point where we wish to find σ_{zz}. Let us put $b = \sqrt{x^2 + z^2}$ and introduce the angle θ shown in the figure. Then eq. (4.26) can be written

$$d\sigma_{zz} = \frac{2p_oz^3dx}{\pi b^4} = \frac{2p_o}{\pi}\cos^2\theta\, d\theta$$

and the stress increase σ_{zz} due to the strip load is found by integrating

$$\sigma_{zz} = \frac{2p_o}{\pi} \int_{\theta_1}^{\theta_2} \cos^2\theta\, d\theta$$

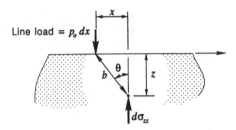

Figure 4.26 Geometry for finding stress increment due to line load.

where the angles θ_1 and θ_2 are illustrated in Figure 4.27. Carrying out the integration we find

$$\sigma_{zz} = \frac{p_o}{\pi} \left[\!\!\left[\theta + \frac{1}{2}\sin 2\theta \right]\!\!\right]_{\theta_1}^{\theta_2}. \tag{4.27}$$

With a similar analysis we can show that

$$\sigma_{xx} = \frac{p_o}{\pi} \left[\!\!\left[\theta - \frac{1}{2}\sin 2\theta \right]\!\!\right]_{\theta_1}^{\theta_2} \tag{4.28}$$

and

$$\sigma_{xz} = \frac{p_o}{\pi} \left[\!\!\left[\sin^2 \theta \right]\!\!\right]_{\theta_1}^{\theta_2}. \tag{4.29}$$

It is also easy to show that $\sigma_{yy} = \nu(\sigma_{xx} + \sigma_{zz})$.

We see from the last three equations that the stress field is completely specified by p_o and the two angles θ_1 and θ_2. We can use Mohr's circle with these expressions to find the principal stresses. They turn out to be

$$\sigma_1 = \frac{p_o}{\pi}(\Theta + \sin\Theta) \quad \text{and} \quad \sigma_3 = \frac{p_o}{\pi}(\Theta - \sin\Theta) \tag{4.30}$$

where $\Theta = \theta_2 - \theta_1$. The intermediate principal stress σ_2 is given by $\nu(\sigma_1 + \sigma_3)$. As (4.30) makes clear, the principal stresses depend only on the angle subtended at the point by the width of the strip. This implies that the principal stresses will be constant on any circle that passes through the edges of the

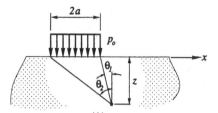

Figure 4.27 Geometry for finding stress due to strip load.

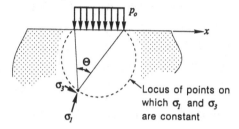

Figure 4.28 Principal stresses due to strip load.

strip load, as shown in Figure 4.28. It can also be shown that the direction of σ_1 points toward the highest point on this circle.

Before we turn our attention to displacements, return to eq. (4.27) and examine the vertical stress in a little more detail. Specifically, we can ask, what happens to σ_{zz} if we consider a point on the half-space surface? First, consider a point outside the strip load. As the point approaches the half-space surface, we see that θ_1 and θ_2 both approach $\pi/2$, and when we evaluate (4.27) we find $\sigma_{zz} = 0$, exactly as expected. Next, consider a point beneath the strip load. Now $\theta_2 = \pi/2$, but $\theta_1 = -\pi/2$. When we evaluate (4.27) the $\sin 2\theta$ term vanishes and we find $\sigma_{zz} = p_o$, again as we expect. Finally, consider a point exactly under the edge of the strip load. For the right-hand edge in Figure 4.27, we would have $\theta_2 = \pi/2$ and $\theta_1 = 0$. Evaluating (4.27) we find $\sigma_{zz} = p_o/2$. Is this what we expect? It is, because if we were to place two similar strip loads side by side, we could use superposition to add their effects, and at the point where they meet, the contribution from each would be $p_o/2$.

Now we can turn to displacements and try to find the settlements caused by the strip load. We did not work out the displacement field for Flamant's problem in Chapter 3; we will do so now. Consider the geometry shown in

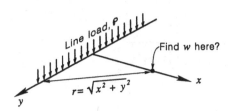

Figure 4.29 Geometry for finding surface settlement due to strip loads.

Figure 4.29. The contribution to the surface settlement w due to the "point load" $\wp\,dy$ will be given by Boussinesq's solution

$$dw = \frac{(\wp\,dy)(1 - \nu)}{2\pi Gr}.$$

To find w we need to integrate over the line

$$w = 2 \int_0^\infty \frac{\wp(1 - \nu)}{2\pi Gr}\,dy.$$

If we introduce the angle $\theta = \tan^{-1}(y/x)$, this can also be written as

$$w = \frac{\wp(1 - \nu)}{\pi G} \int_0^{\pi/2} \sec\theta\,d\theta.$$

When we integrate, we find

$$w = \frac{\wp(1 - \nu)}{\pi G} \left[\!\!\left[\ell_n(\sec\theta + \tan\theta) \right]\!\!\right]_0^{\pi/2}$$

or

$$w = \frac{\wp(1 - \nu)}{\pi G} \left[\!\!\left[\ell_n(y + \sqrt{x^2 + y^2}) \right]\!\!\right]_{y=0}^{y=\infty} \tag{4.31}$$

From either of these last two equations, we see that w is infinite for all values of x. This is the peculiarity we mentioned earlier. Every point on the half-space surface suffers an infinite settlement. If the line load has any finite length, this problem doesn't arise. The settlements will be finite for all points except those beneath the line load itself (and we expect them to be infinite there). But if we let the line load become longer and longer, then the settlements increase in proportion to the natural logarithm of the length of the load, and if the load becomes infinite, so too do the settlements. This may seem a serious problem, but we must bear in mind that infinitely long line loads are not found in nature. They are a mathematical artifice. If they lead us to an unexpected result, we must consider that result in the context of the somewhat artificial problem we have set.

Sometimes engineers attempt to avoid the problem of infinite settlement in plane-strain conditions by introducing the *relative* settlement Δw between two points on the half-space surface. For two points x_1 and x_2, say, the relative settlement at x_1 relative to x_2 is defined by

$$\Delta w = w(x_1) - w(x_2).$$

Of course, both $w(x_1)$ and $w(x_2)$ are infinite, but their difference is not. If we evaluate (4.31) at x_1 and at x_2 and subtract, we find

$$\Delta w = \frac{\rho(1 - \nu)}{\pi G} \, \ell n \left(\frac{x_2}{x_1} \right). \tag{4.32}$$

For the strip load, absolute settlements are also infinite, but relative settlements are not. If we apply (4.32) to the strip load illustrated in Figure 4.30, we find

$$\Delta w = \frac{p_o(1 - \nu)}{\pi G} \left\{ x_2 \ell n \left(\frac{x_2 + 2a}{x_2} \right) - x_1 \ell n \left(\frac{x_1 + 2a}{x_1} \right) + a \ell n \left(\frac{x_2 + 2a}{x_1 + 2a} \right) \right\}.$$

This result is interesting but may not be especially useful since we can find any value of Δw by choosing the point x_2 at greater and greater distances from the strip load. Equation (4.32) will be useful only when *relative* displacements are required for some particular reason. If we wish to know the absolute amount of settlement at some point on the half-space surface, then we must take account of the true length of the loaded region and use the expressions derived earlier for rectangular loads.

Next, we can pose the question of how an infinitely long rigid footing will behave on an elastic half-space. We can immediately deduce the absolute settlements will be infinite and turn attention to relative displacements. The relative displacement profile will be as shown in Figure 4.31, where settlements are measured relative to some point outside the footing.

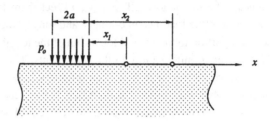

Figure 4.30 Geometry for finding relative settlement due to strip load.

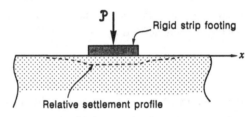

Figure 4.31 Rigid strip footing.

The relative settlement beneath the footing will be a constant that we will call δ. To solve the problem, we need to find the contact stress distribution $p(x)$ that gives a constant displacement beneath the footing and that equilibrates the applied load

$$\int_{-a}^{a} p(x)dx = \mathcal{P}. \tag{4.33}$$

Note here that $\mathcal{P}$ is the load per unit length of footing, and we assume $\mathcal{P}$ is constant, otherwise plane-strain conditions would not apply.

We might anticipate that the contact stress may become singular as we approach the footing edge, just as was the case for the rigid circular footing. The correct contact stress distribution, discovered by M. Sadowsky in 1928, is given by

$$p(x) = \frac{\mathcal{P}}{\pi\sqrt{a^2 - x^2}} \quad \text{for } -a < x < a. \tag{4.34}$$

We see that $p(x) \rightarrow \infty$ as x approaches $\pm a$, and it is easy to verify that equilibrium is preserved by substituting (4.34) in (4.33) and integrating. The rel-

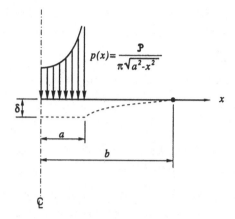

$$p(x) = \frac{\mathcal{P}}{\pi\sqrt{a^2 - x^2}}$$

Figure 4.32 Contact stress and settlement profiles for the rigid strip footing.

ative settlement δ, relative to a point located a distance b from the footing center (see Figure 4.32) can be found by using eq. (4.32) with $\mathcal{P}$ replaced by the "line load" $p(x)dx$ and integrating. We find

$$\delta = \frac{\mathcal{P}(1 - \nu)}{\pi G} \, \ell n \left(\frac{b}{a} + \sqrt{\frac{b^2}{a^2} - 1} \right). \tag{4.35}$$

Take care here to note that δ is the *relative* settlement of the rigid strip footing, relative to the settlement at point b.

Comments made earlier with regard to observed stress distributions beneath rigid footings on cohesionless and cohesive soils will apply here as well. For a rigid strip footing resting on the surface of a deposit of cohesionless sand, a roughly parabolic stress distribution (Figure 4.18) would be the expected result. On a clay soil, an inverted parabolic distribution (Figure 4.19) would be expected. The reasoning behind these expectations was explained in Section 4.5. It should be clear that we can use Flamant's solution and eq. (4.32) to work out the details of the stress field and the relative settlements for other problems involving various stress distributions that obey plane-strain conditions.

4.7 Settlements in layered soils

The next question we want to consider is this: What if the soil profile at some site is not homogeneous? More likely than not, some degree of inhomogene-

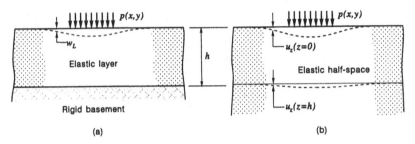

Figure 4.33 The problem of an elastic layer on a rigid base approximated by the local compression in an elastic half-space.

ity will exist in most natural soil profiles. The most likely profile is layered. Transported soils will almost always exhibit layering. The natural processes of transportation and deposition by air and water lead automatically to layering on a variety of scales. Residual soils tend to be more uniform, or at least to change more gradually, but layering may be evident in distinct weathering horizons. Other types of inhomogeneity will also be present in the form of lateral variations and occasional inclusions and lenses and even voids in rare instances. These sorts of inhomogeneities are beyond our scope here, but regular layering is not. Fundamental solutions for layered elastic media were investigated by Burmister, and were briefly mentioned in Chapter 3. These solutions are exact but complex. Instead of attempting to use them, there is a very simple way to approximate the settlements in a layered soil profile that we will explain.

We can begin by considering the simplest layered problem, a single elastic layer resting on a rigid basement as shown in Figure 4.33(a). The layer has thickness h, and we wish to find the surface settlement w_L due to a specified applied stress $p(x,y)$. Figure 4.33(b) illustrates how we might approach this problem. Suppose we have an elastic half-space with the same elastic properties as the elastic layer. Consider the response of the half-space if we apply the same loading which acts on the layer. In particular, consider the surface settlement $u_z(z = 0)$ and the vertical displacement $u_z(z = h)$. We might expect the relative displacement $u_z(z = 0) - u_z(z = h)$ to be a reasonable approximation for w_L since it represents the local compression of the half-space above the depth h. It turns out this is a good approximation as long as the layer thickness h is not too small when compared with the lateral extent of the loaded region.

To illustrate this method, suppose we have a uniform circular load of intensity p_o and radius a acting on the surface of an elastic layer over a rigid basement. We apply the same loading to a half-space with similar elastic constants. The surface settlement at the center of the loaded region is given by eq. (4.3)

$$u_z(z = 0) = \frac{p_o a(1 - \nu)}{G}$$

and the vertical displacement at $r = 0$, $z = h$ is given by eq. (4.10)

$$u_z(z = h) = \frac{p_o}{2G} \left\{ 2(1 - \nu)(\sqrt{a^2 + h^2} - h) - h\left(\frac{h}{\sqrt{a^2 + h^2}} - 1\right) \right\}.$$

So our approximation for the layer settlement w_L at the center of the loaded area is

$$w_L = u_z(z = 0) - u_z(z = h)$$

$$= \frac{p_o}{2G} \left\{ 2(1 - \nu)(a + h - \sqrt{a^2 + h^2}) - h\left(1 - \frac{h}{\sqrt{a^2 + h^2}}\right) \right\}. \quad (4.36)$$

Note that the limiting behavior of this expression as $h \to 0$ or as $h \to \infty$ is what we would expect to find. For intermediate values of h, (4.36) gives a reasonably good approximation for the exact settlement found using Burmister's solution. The maximum degree of error is on the order of 10% except for cases where the ratio h/a is quite small. See the article by Poulos (1967) for further details.

Obviously, we can generalize this approach to consider multiply layered sites. For example, if we have a soft soil layer lying on a deep deposit of stiffer soil as shown in Figure 4.34, we can approximate the settlement w_L as follows:

$$w_L = u_z^{(1)}(z = 0) - u_z^{(1)}(z = h) + u_z^{(2)}(z = h)$$

where $u_z^{(1)}$ are displacements calculated for an elastic half-space with elastic properties ν_1 and G_1, while $u_z^{(2)}$ is the half-space displacement using elastic constants ν_2 and G_2. The relative displacement $u_z^{(1)}(z = 0) - u_z^{(1)}(z = h)$ approximates the compression in the soft layer, while the third term $u_z^{(2)}(z = h)$ approximates the contribution to the settlement from the compression of the underlying soil. If more layers are present, we can approximate the local compression in each using half-space calculations and then add them all together to approximate the total settlement. For the circular-load problem these

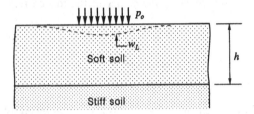

Figure 4.34 Typical layered-soil problem.

calculations are quite simple. For other load geometries, finding the vertical displacement at depth may lead to difficult integrations, but in most common cases tables and graphical solutions can be used to find the displacement, at least to an accuracy consistent with the degree of approximation desired. The book by Poulos and Davis cited in Chapter 1 is a valuable source for these calculations.

4.8 Consolidation and settlement

In any soil, time may be required for settlements to occur. This is especially true, of course, for fine-grained, fully-saturated soils. In these soils, the time dependence results from the process of *consolidation*, familiar to all students of soil mechanics. The analysis of consolidation achieved by Karl Terzaghi in 1925 has become a cornerstone of geotechnical engineering.

An important element of Terzaghi's theory was the assumption of one-dimensional motions for both the solid-particle skeleton and the pore fluid. In Terzaghi's theory, the soil skeleton is assumed to undergo uniaxial compression (discussed in Section 2.9) where only one strain component (the vertical strain ϵ_{zz}) is nonzero. Only one elastic constant is needed to fully describe the solid-skeleton behavior, a constant similar to that in eq. (2.36). In Terzaghi's theory, the pore fluid is also assumed to move only in the vertical direction. Its motion is governed by Darcy's law, and the coefficient of permeability is a new parameter required to fully describe the solid-skeleton/pore-water system. Settlement occurs as the pore fluid is squeezed from the matrix of solid particles. Applied loads on the soil surface are initially supported by hydrostatic stress in the pore fluid, but flow toward drainage boundaries removes pore fluid from the soil, and the load is gradually transferred to the solid-particle skeleton as compression occurs. These concepts are discussed in every elementary text on soil mechanics.

In Terzaghi's theory, there is no mechanism to permit *immediate settlement* to occur. By immediate settlement, we mean deformations that occur simultaneously with the application of external loads. Terzaghi assumed the individual solid particles and the pore fluid were both incompressible. This is a good approximation, of course, since the compressibility of nearly all soil-forming minerals is much smaller than that of water, and both are far less compressible than the solid-particle matrix in nearly any real soil. We can safely assume (as Terzaghi did) that all deformation occurs because of localized rearrangement of solid particles leading to a decrease in pore volume. But in Terzaghi's one-dimensional theory, no decrease in pore volume can occur until flow begins, and hence no deformation can occur simultaneously with load application. This is all a result of the one-dimensional nature of Terzaghi's theory.

At this point one might ask, why worry about all this? The one-dimensional theory is simple and easy to use and yields a vast amount of understanding of soil behavior. Why not just stick with it? The answer to that question is twofold. First, we observe immediate settlements that are often quite important. Second, the rate of settlement predicted by Terzaghi's theory may be significantly slower than will actually occur in a three-dimensional, real soil application.

The term *three-dimensional* is the key to understanding. Immediate settlements occur because the soil can deform in three dimensions rather than just one, and long-term settlements will occur at a rate that depends on three-dimensional flow patterns within the soil deposit. A complete three-dimensional theory of consolidation and settlement is available. It was derived by the Belgian mechanicist Maurice Biot in 1940, and it correctly accounts for immediate and long-term settlements. It is based on the equations of elasticity to describe the solid particle-skeleton deformation combined with Darcy's law to govern the flow of the pore fluid. It is beyond our scope here to describe Biot's theory in any detail. We can, however, make some simple observations about three-dimensional effects in consolidation.

First, we confine ourselves to discussion of fully saturated soils. We assume the soil pores are completely filled with water. We can also assume that both the individual solid particles and the pore fluid are incompressible. Then, unless flow can occur, the soil itself is incompressible. We know how to characterize an incompressible material. We discussed the subject in Section 2.11. All we need to do is set ν equal to 0.5. Both the bulk modulus and the Lamé constant λ become infinite and $E = 3G$. Immediate settlements are simply the deformations of this incompressible soil, and we can use all of the developments of this chapter to find immediate settlements simply by using the incompressible or undrained, elastic constants.

Following the immediate settlements, further deformations will occur due to consolidation effects. We can refer to these as consolidation settlements. Added together, the immediate and consolidation settlements give the total or final settlements associated with the applied loads. We can find these final settlements by means of elasticity solutions using values of elastic constants appropriate to the solid-particle matrix of the soil, the values often called the elastic constants for drained conditions.

If we use ν and G as our elastic constants, as we have done throughout this chapter, then the only difference between the immediate, undrained solution and the final, drained solution lies in the value of ν. The shear modulus G is not affected by the presence of the pore fluid, and it remains the same in either case. The value of Poisson's ratio does change, from the undrained value of 0.5 to some lesser value appropriate to the fully drained soil skeleton. In an intuitive way, we can look on the three-dimensional consolidation process as a relaxation of Poisson's ratio from 0.5 to some smaller value.

To summarize these ideas, consider a uniform circular load on the surface of a deep homogeneous deposit of fully saturated clay. Let the shear modulus and Poisson's ratio (for drained conditions) be G and ν. The immediate settlement that occurs simultaneously with application of the circular load is given by (4.3) with ν set equal to 0.5.

$$\text{Immediate settlement} = w_i = \frac{p_o a}{2G} \qquad (4.37)$$

where p_o is the intensity of the load and a the radius of the loaded region. The final settlement is also found from (4.3) but with Poisson's ratio taking its drained value

$$\text{Final settlement} = w_f = \frac{p_o a(1 - \nu)}{G}. \qquad (4.38)$$

The consolidation settlement is the difference between these

$$\text{Consolidation settlement} = w_c = w_f - w_i$$

$$= \frac{p_o a(1 - 2\nu)}{2G}.$$

Comparing (4.37) with (4.38) we can see what fraction of the total settlement occurs immediately

$$\frac{w_i}{w_f} = \frac{1}{2(1 - \nu)}.$$

Recalling that ν will always lie between zero and 0.5, we see the immediate settlements will account for at least half the total settlement that occurs. Measurements of immediate settlement are therefore very interesting. Within the context of elastic behavior, we can be confident in predicting the final settlement will be no greater than twice the immediate settlement. This is a prediction that can be made without any site investigation or soil testing whatever. The qualification concerning elastic behavior is important, however.

Finally, we can ask how much time will be required for the consolidation settlements to fully develop? A number of investigators have studied a range of problems involving different loading geometries, different drainage conditions, and the effects of layered soils. These problems are beyond the scope of this book, but in general we might expect three-dimensional effects to result in faster consolidation times than those predicted by Terzaghi's one-dimensional theory. In the three-dimensional problem, the pore fluid can move horizontally as well as vertically and this added dimension suggests more rapid consolidation. Generally this is the case, especially in situations where the characteristic dimensions of the loaded region are significantly less than the depth of the soil deposit.

4.9 Applications to in situ testing

In this section, we will look back at some of the results mentioned in Section 2.10, where we discussed field tests. Recall the plate-load test illustrated in Figure 2.14. We presented in eq. (2.38) a relationship between the load $\mathcal{P}$, the deflection δ, and the elastic constants, for the case of a rigid circular plate. Equation (2.38) results from the analysis of the rigid foundation we discussed in Section 4.5. In fact, eq. (2.38) is exactly the same as eq. (4.25). Results for the rough circular plate in eq. (2.39) and the square plate in eq. (2.40) similarly follow directly from rigid foundation analyses.

The screw-plate test also discussed in Section 2.10 may similarly be analyzed. We appeal to Kelvin's solution discussed in Section 3.3. We assume the plate is embedded in an infinite elastic body as shown in Figure 4.35, and we can represent the screw plate by an assumed load distribution or as a rigid plate displaced by an amount δ. The simplest analysis is to assume a uniform stress applied over the circular region that has the plate radius a. Referring back to Kelvin's problem, recall that we used a point load of magnitude $2P$. In the context of the circular applied stress, we will replace $2P$ by $p_o r d\theta dr$

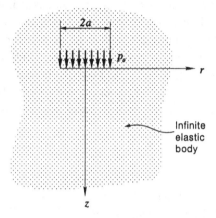

Figure 4.35 Analysis of the screw plate: uniform circular load embedded in an infinite elastic body.

and then integrate (3.22) to find the vertical displacement of the plate. At the plate center we have $z = 0$ and hence

$$u_z = \int_0^{2\pi} \int_0^a \frac{(p_o r \, dr \, d\theta)}{16\pi G(1 - \nu)} \left[\frac{2(1 - 2\nu)}{r} + \frac{1}{r} \right].$$

Carrying out the integrations we find

$$u_z = \frac{p_o a(3 - 4\nu)}{8G(1 - \nu)}.$$

The stress p_o can be replaced by the total applied load $\mathcal{P}$ divided by the plate area to have

$$u_z = \frac{\mathcal{P}(3 - 4\nu)}{8\pi a G(1 - \nu)}. \tag{4.39}$$

For a rigid plate, fully bonded to the surrounding material, the solution is

$$u_z = \frac{\mathcal{P}(3 - 4\nu)}{32 a G(1 - \nu)}. \tag{4.40}$$

Other solutions for the partially bonded rigid plate and a flexible plate are also available.

To conclude this section, we will briefly discuss the pressuremeter test. Recall from Chapter 2 that the pressuremeter employs a cylindrical membrane to be inflated inside a borehole. Both the inflation pressure and the displacement of the bore wall are measured. This problem, and other similar ones, fall into a category called cavity-expansion problems. They have been widely studied in regard to a number of different applications. We will just consider the elastic cylindrical cavity problem here.

We can begin by idealizing the pressuremeter geometry. Assume that the bore is infinitely long and its inside wall is subjected to a uniform applied stress p_o. Under these conditions we have a plane-strain problem. If the z-axis is aligned with the center of the bore, then every plane surface defined by z-constant behaves exactly the same. Looking at Figure 4.36, it is physically reasonable to assume only one nonzero displacement u_r, which will be only a function of r. The strain displacement relations for cylindrical geometry are obtained by combining (1.3) with (3.2). We find only two non-zero strains are possible

$$\epsilon_{rr} = -\frac{du_r}{dr} \quad \text{and} \quad \epsilon_{\theta\theta} = -\frac{u_r}{r}. \tag{4.41}$$

The negative signs appear here because we want compressive strains to be positive. If we now appeal to Hooke's law, we find the two non-zero strains can produce only three nonzero stresses: σ_{rr}, $\sigma_{\theta\theta}$, and σ_{zz}. Moreover, since $\epsilon_{zz} = 0$ we can conclude $\sigma_{zz} = \nu(\sigma_{rr} + \sigma_{\theta\theta})$, so only two stresses are needed.

Next, we know we need to satisfy the equations of equilibrium. For cylindrical geometry, the equations are written out in (3.14). Only one of those three equations is not trivially satisfied

$$\frac{d\sigma_{rr}}{dr} + \frac{1}{r}(\sigma_{rr} - \sigma_{\theta\theta}) = 0. \tag{4.42}$$

We also must specify appropriate boundary conditions. If the borehole radius is a, then the radial stress must equilibrate the applied pressure at that point

$$\sigma_{rr}(r = a) = p_o. \tag{4.43}$$

A second boundary condition is also needed. We find it by moving as far from the borehole as possible:

$$\sigma_{rr}(r \to \infty) = 0. \tag{4.44}$$

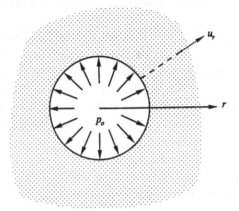

Figure 4.36 Geometry for the pressuremeter problem: expansion of a cylindrical cavity.

These equations now fully specify the problem.

To solve this problem, we can use (4.41) in Hooke's law to have

$$\sigma_{rr} = \lambda\left(-\frac{du_r}{dr} - \frac{u_r}{r}\right) + 2G\left(-\frac{du_r}{dr}\right)$$

$$\sigma_{\theta\theta} = \lambda\left(-\frac{du_r}{dr} - \frac{u_r}{r}\right) + 2G\left(-\frac{u_r}{r}\right)$$

and using these in the equilibrium equation (4.42) we find

$$(\lambda + 2G)\left(\frac{d^2u_r}{dr^2} + \frac{1}{r}\frac{du_r}{dr} - \frac{u_r}{r^2}\right) = 0.$$

Since $\lambda+2G$ cannot be zero, the term in brackets must vanish. This gives

$$\frac{d}{dr}\left(\frac{du_r}{dr} + \frac{u_r}{r}\right) = 0$$

which has the general solution

$$u_r = Ar + \frac{B}{r} \tag{4.45}$$

where A and B are constants to be specified.

To complete the solution, we employ our boundary conditions. First, we use (4.45) in (4.41) to find strains

$$\epsilon_{rr} = -A + \frac{B}{r^2} \quad \text{and} \quad \epsilon_{\theta\theta} = -A - \frac{B}{r^2}.$$

Using these in Hooke's law, we obtain the radial stress field

$$\sigma_{rr} = (\lambda + G)(-2A) + 2G\left(\frac{B}{r^2}\right).$$

The boundary condition (4.44) shows A must be zero. The boundary condition (4.43) then gives

$$B = \frac{p_o a^2}{2G}.$$

Finally, from (4.45) we now have

$$u_r(r) = \frac{p_o a^2}{2Gr}.$$

Evaluating this expression at the borehole wall, we can write an equation for the shear modulus that depends only on displacement and the pressure in the bore

$$2G = \frac{p_o a}{\Delta}. \tag{4.46}$$

Here $\Delta = u_r(r = a)$ is the radial displacement at the bore wall. It is this equation that is used in conjunction with pressuremeter measurements to estimate the shear modulus.

4.10 Gravity stresses in earth structures

To complete this chapter, we will consider an interesting problem concerning gravity stresses solved by L. E. Goodman and C. B. Brown in 1963. Goodman and Brown were concerned about the way an earth structure such as an embankment is constructed and how the method of construction may affect the gravity stresses inside the finished embankment. They noted that an embankment is built up in horizontal layers, and the gravity stresses must evolve in such a way to reflect the addition of each layer. They also noted that these evolving stresses might well not be the same as those we would find if gravity were suddenly "switched on" after the embankment was fully built. In the case of a horizontal ground surface, such as in Figure 3.18, it makes no difference since the evolving stresses will always be the same regardless of how the soil may have been deposited. But if the ground surface is not flat, this is not the case.

Goodman and Brown considered the situation illustrated in Figure 4.37. We see part of an embankment: the horizontal top surface and one sloping surface. We do not see the other sloping surface or the base of the embankment; to let these enter the problem would make things too difficult. Nevertheless, we can rely on Saint-Venant's principle to suggest that the stress field will not be strongly affected by the embankment geometry away from the region of interest. The angle α measures the side slope. It can vary between $\pi/2$, which would correspond to a vertical slope, and π for a completely horizontal surface.

When embankments are constructed, layers of soil are placed one on top of the next and compacted. As any particular layer is added, it has the effect of placing additional stress on the soil already in place. The increment of added stress is the thickness of the layer Δ multiplied by its unit weight ρg. The situation is illustrated in Figure 4.38. We see the embankment partially constructed and a uniform applied stress $\rho g \Delta$ acting on the upper surface. Goodman and Brown foresaw that if they could solve the problem in Figure 4.38, then they could continue adding layers until the embankment reached its final shape. If the layers are thin, then the addition of layers becomes an integration. The final stress state, at point A for example, is the cumulative effect of the weights of each of the layers beginning with the layer immediately above A. The interesting point is this: Body forces play no role in the solution. The stress at A is found by solving the elasticity problem without body forces. Only the applied stress on the upper surface affects the stress at A. Yet after all the layers are in place, the stress at A is the gravity stress. Its value may be different, however, from the stress we would find by building the embankment with "weightless" soil and switching gravity on in the final embankment shape.

In Figure 4.38, we have introduced a secondary coordinate system X, Z, which lies at the top of the present embankment configuration. This coordinate system can move as each layer is added until finally it coincides with the x, z system when the embankment is complete. We would now like to solve the traction boundary value problem with the uniform applied stress $\rho g \Delta$ act-

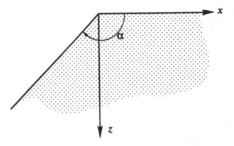

Figure 4.37 An earth embankment.

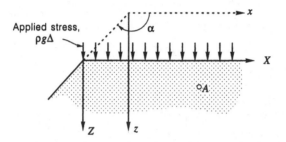

Figure 4.38 Stress increment due to addition of a layer of soil of thickness Δ.

ing on the surface $X \geq 0$, $Z = 0$. We will assume plane-strain conditions. Compressive stresses are taken as positive.

The problem is redrawn in Figure 4.39. Goodman and Brown attacked the problem with an Airy stress function. They used the following stress function (Carothers, 1913)

$$\Phi(X,Z) = \frac{\rho g \Delta}{2(\alpha - \tan\alpha)} [(\alpha - \theta)(X^2 + Z^2) + XZ - X^2\tan\alpha] \quad (4.47)$$

where $\theta = \tan^{-1}(Z/X)$ is the angle shown on Figure 4.39. Recalling the Airy stress function from Section 3.7, we see that (4.47) must satisfy the biharmonic equation. The polynomial terms in X and Z clearly will work. The term $\theta(X^2 + Z^2)$ is a little more complex, but we investigated a similar form in question 3.10 and found that it worked as well. The stress field associated with (4.47) is given by eqs (3.43), where we note that V is zero. Body forces have no role in this solution. Carrying out the differentiations we find

$$\sigma_{XX} = \frac{\partial^2\Phi}{\partial Z^2} = \frac{\rho g \Delta}{\alpha - \tan\alpha} (\alpha - \theta - \sin\theta \cos\theta)$$

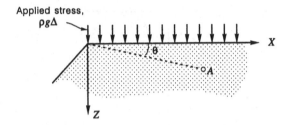

Figure 4.39 Partially completed embankment, incremental solution.

$$\sigma_{ZZ} = \frac{\partial^2 \Phi}{\partial X^2} = \frac{\rho g \Delta}{\alpha - \tan\alpha} (\alpha - \theta - \tan\alpha + \sin\theta \cos\theta)$$

$$\sigma_{XZ} = \frac{-\partial^2 \Phi}{\partial X \partial Z} = \frac{-\rho g \Delta}{\alpha - \tan\alpha} \sin^2\theta. \tag{4.48}$$

These stresses will satisfy the plane-strain equilibrium equations (3.34) without body forces. They will also satisfy the boundary conditions

$$\text{on} \quad Z = 0, X \geq 0: \begin{cases} \sigma_{ZZ} = \rho g \Delta \\ \quad\quad \sigma_{XZ} = 0 \end{cases}$$

$$\text{on} \quad Z = X \tan\alpha, X < 0: \begin{cases} \sigma_{XZ} - \sigma_{XX} \tan\alpha = 0 \\ \sigma_{ZZ} - \sigma_{XZ} \tan\alpha = 0 \end{cases}.$$

Note that on the side slope where $Z = X \tan\alpha$ we have $\theta = \alpha$.

In order to extend this solution to the completed embankment, Goodman and Brown let Δ become small (so that it approached dZ) and then integrated the incremental stress equations upward to the completed embankment height. This gave the following stresses:

$$\sigma_{xx} = \frac{\rho g}{\alpha - \tan\alpha} \left\{ z(\alpha - \sin\alpha \cos\alpha) \right.$$

$$- 2\sin\alpha \cos^2\alpha (x \sin\alpha - z \cos\alpha) \ell n \left(\frac{\sin(\alpha - \theta)}{\sin\alpha} \right)$$

$$\left. - \theta \sin\alpha [z \sin\alpha (1 + 2\cos^2\alpha) + x \cos\alpha (1 - 2\sin^2\alpha)] \right\}$$

$$\sigma_{zz} = \rho g z - \frac{\rho g}{\alpha - \tan\alpha} \left\{ -z \sin\alpha \cos\alpha \right.$$

$$+ 2\sin^3\alpha(x \sin\alpha - z \cos\alpha)\ln\left(\frac{\sin(\alpha - \theta)}{\sin\alpha}\right)$$

$$\left. + \theta\sin\alpha[z \sin\alpha(1 - 2\cos^2\alpha) + x \cos\alpha(1 + 2\sin^2\alpha)]\right\}$$

$$\sigma_{xz} = \frac{-\rho g}{\alpha - \tan\alpha} \left\{ z \sin^2\alpha \right. \tag{4.49}$$

$$+ 2\sin^2\alpha\cos\alpha(x \sin\alpha - z \cos\alpha)\ln\left(\frac{\sin(\alpha - \theta)}{\sin\alpha}\right)$$

$$\left. + \theta \sin\alpha(1 - 2\sin^2\alpha)(x \sin\alpha - z \cos\alpha)\right\}.$$

Here θ represents $\tan^{-1}(z/x)$ in the final embankment geometry.

The stresses in (4.49) will now obey the equilibrium equations (3.34) with the body force f_z *not zero but equal to* $-\rho g$. This occurs despite the absence of body forces in the incremental solution. Equations (4.49) will also satisfy the boundary equations for the completed embankment

$$\text{on} \quad z = 0, x \geq 0: \begin{cases} \sigma_{zz} = 0 \\ \sigma_{xz} = 0 \end{cases}$$

$$\text{on} \quad z = x \tan\alpha, x < 0: \begin{cases} \sigma_{xz} - \sigma_{xx} \tan\alpha = 0 \\ \sigma_{zz} - \sigma_{xz} \tan\alpha = 0 \end{cases}.$$

Equations (4.49) will not, however, satisfy the plane-strain compatibility equation (3.42) even though the incremental stresses do. This peculiar phenomenon results because each additional layer of soil in the incremental solution permits some horizontal motion or slip at the present upper surface of the embankment. That slip is then locked into the embankment as further layers are placed, and overall compatibility cannot be satisfied.

In their paper, Goodman and Brown plotted equations (4.49) in dimensionless form for several values of slope angle α. They also considered the implications of the stresses on the stability of the embankment. Rather than reproduce their findings, we will consider just one aspect of their solution. We will investigate the stresses that act on a surface parallel to the side-slope surface as shown in Figure 4.40. We let h be the vertical depth to the surface and note that for any point on the surface

$$x \sin\alpha - z \cos\alpha = -h \cos\alpha$$

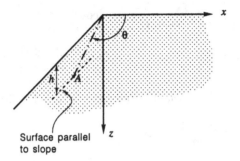

Figure 4.40 The completed embankment.

and

$$z \sin\alpha + x \cos\alpha = \frac{z - h}{\sin\alpha} + h \sin\alpha. \tag{4.50}$$

The unit normal vector to the surface has components

$$\hat{n} = \begin{bmatrix} \sin\alpha \\ -\cos\alpha \end{bmatrix}$$

and the components of the traction vector on the surface are

$$T = \sigma^T \hat{n} = \begin{bmatrix} \sigma_{xx} \sin\sigma - \sigma_{xz} \cos\alpha \\ \sigma_{xz} \sin\alpha - \sigma_{zz} \cos\alpha \end{bmatrix}.$$

Now let σ and τ denote the components of T normal and tangential to the surface

$$\sigma = T_x \sin\alpha - T_z \cos\alpha = \sigma_{xx} \sin^2\alpha - 2\sigma_{xz} \sin\alpha \cos\alpha + \sigma_{zz} \cos^2\alpha$$

$$\tau = -T_x \cos\alpha - T_z \sin\alpha = (\sigma_{zz} - \sigma_{xx}) \sin\alpha \cos\alpha + \sigma_{xz} (\cos^2\alpha - \sin^2\alpha). \tag{4.51}$$

Note that the cosine is negative for α between $\pi/2$ and π. If we now use (4.50) in (4.49) and then use the resulting equations for the stresses in (4.51) we find

$$\sigma = \frac{\rho g h}{\alpha - \tan\alpha} \left[\frac{z}{h}(\alpha - \theta) + \theta \cos^2\alpha \right]$$

$$\tau = \frac{\rho g h}{\alpha - \tan\alpha} [\theta \sin\theta \cos\theta]. \tag{4.52}$$

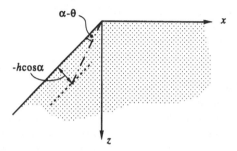

Figure 4.41 Geometry of surface parallel to embankment surface.

What if we now consider a point well below the embankment top? Then θ will be close to α. The geometry is shown in Figure 4.41. We see that

$$\alpha - \theta = \frac{-h \cos\alpha}{(x^2 + z^2)^{1/2}}$$

so that

$$\frac{z}{h}(\alpha - \theta) = -\cos\alpha \sin\theta.$$

If we use this in (4.52) and then let $\theta \to \alpha$ we find

$$\sigma = \rho g h \cos^2\alpha$$

$$\tau = \rho g h \sin\alpha \cos\alpha \left(1 + \frac{\tan\alpha}{\alpha - \tan\alpha}\right). \qquad (4.53)$$

We have written τ in this awkward way to make a particular point. We can compare σ and τ here with the stress components σ_{zz} and σ_{xz} for the long slope found in Section 3.7. The geometry in Section 3.7 was shown in Figure 3.19. There the z axis was directed normal to the slope surface, and we see that σ_{zz} and σ_{xz} in (3.51) are equivalent to σ and τ in (4.53), provided we set $\sin\alpha = \sin\beta$, $\cos\alpha = -\cos\beta$, and $z = -h \cos\alpha$. When we do that we see that σ agrees exactly with σ_{zz} in (3.51), but τ differs from σ_{xz} by the term $\tan\alpha/(\alpha - \tan\alpha)$. This difference is the result of building up the embankment in layers.

By solving this problem in the way Goodman and Brown suggest, we have found a stress component different from the problem where gravity is simply

Applications of fundamental solutions

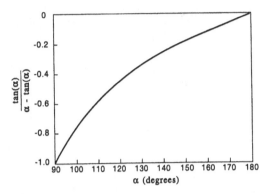

Figure 4.42 Variation in shear stress on surface parallel to embankment surface.

switched on. Moreover, it is a particularly interesting stress component, since the surface parallel to the slope is important from the standpoint of stability, and it is the shear stress component that is altered. A graph of the term $\tan\alpha/(\alpha - \tan\alpha)$ as α varies between $\pi/2$ and π is shown in Figure 4.42. We see that $\tan\alpha/(\alpha - \tan\alpha)$ varies between zero and -1 as the side slope angle changes from horizontal to vertical. From a practical standpoint, it is unlikely the side slope would be steeper than 1:1, otherwise elastic behavior would not be expected. Consequently, for $\alpha > 135°$ the value of $\tan\alpha/(\alpha - \tan\alpha)$ will lie somewhere between about -0.3 and zero. The effect is to reduce the shear stress τ to a value smaller than we would expect based on the conventional analysis of Section 3.7. In a general sense this is clearly a good thing. By building up the embankment in layers we have evidently improved its overall stability.

Exercises

4.1 Verify equation (4.29).

4.2 Given the Cartesian components of stress for the strip load problem in eqs. (4.27), (4.28), and (4.29), derive the expressions for principal stresses in eqs. (4.30).

4.3 For the parabolic stress distribution applied over a circular region on the surface of an elastic half-space given in equation (4.19), find the vertical displacement w at the circumference of the loaded region.

4.4 A uniform stress p_o is applied on the surface of an elastic half-space over a region that is an equilateral triangle with sides of length a as shown below. Find the vertical displacement of the half-space surface at the point marked A.

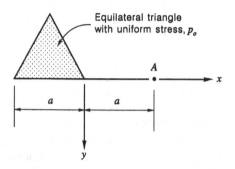

4.5 An infinitely long strip load of width $2b$ with pyramid-shaped stress distribution shown in the figure below acts on the surface of a homogeneous elastic half-space. Determine the vertical component of stress σ_{zz} at the point B shown on the figure.

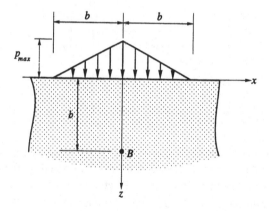

4.6 Estimate the settlement of the ground surface a distance ℓ from an infinite line load of magnitude $\wp$ per unit length for the case of an elastic layer of thickness h resting on rigid basement rock as shown below.

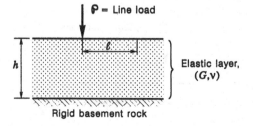

4.7 The figure below shows the plan area of a uniform load p_o acting on the surface of a homogeneous elastic half-space. Find the settlement of the half-space surface at the points marked A, B, and C.

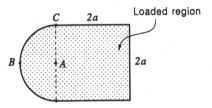

4.8 Suppose the triangular-shaped region shown in the figure below supports a linearly varying load on the surface of a homogeneous elastic half-space. Let p_A, p_B, and p_C be the magnitudes of the applied load at the vertices marked A, B, and C. Determine the vertical settlement of the half-space surface at the vertex A.

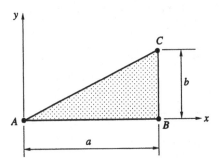

4.9 A line of three circular rigid footings are connected by a rigid cross beam as shown below. The central load F imposed on the cross beam causes it and the three footings to all settle by an amount δ. Use Boussinesq's solution for the rigid circular foundation [eq. (4.25) and the following equation] to estimate the following:

i) the fraction of the load F carried by each footing;

ii) the load-settlement ($F - \delta$) relationship for the beam and footings expressed as a function of G, ν, and a.

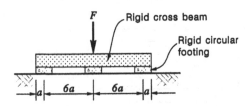

4.10 Modify the strip-load solution from Section 4.6 to obtain the stress field in a half-space that supports a uniform stress extending from $x = 0$ to $x = \infty$ as shown in the figure below. Compare your solution with the solution of Goodman and Brown in equations (4.48) for the special case where $\alpha = \pi$.

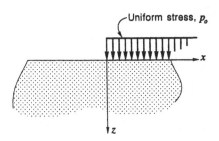

References

Modern works cited in this chapter are:

Biot, M. A., "General theory of three-dimensional consolidation," *Jour. Appl. Phys.*, Vol. 12, pp. 155–164 (1940).
Dempsey, J. P., and Li, H., "A rigid rectangular footing on an elastic layer," *Geotechnique*, Vol. 39, pp. 147–152 (1989).
Gibson, R. E., "The analytical method in soil mechanics," *Geotechnique*, Vol. 24, pp. 115–140 (1974).
Poulos, H. G., "Stresses and displacements in an elastic layer underlain by a rough rigid base," *Geotechnique*, Vol. 17, pp. 378–410 (1967).
Sadowsky, M. A., "Zweidimensionale probleme der elastizitatstheroie," *Z. Angew. Math. Mech.*, Vol. 8, pp. 107–121 (1928).
Terzaghi, K., *Erdbaumechanik auf Bodenphysikalischer Grundlage*, Franz Deuticke, Vienna (1925).
Winkler, E., *Die Lehre von der Elastizitat und Festigkeit*, Dominicus, Prague (1867).

The work on gravity stresses in earth structures is described in

Goodman, L. E., and Brown, C. B., "Dead load stresses and the instability of slopes," *Jour. Soil Mech. Found. Div. ASCE*, Vol. 89, No. SM3, pp. 103–134 (1963).

The solution for the uniform load acting on an elastic wedge (4.47) can be traced to the following works

Lévy, M., "Sur la légitimité de la règle dite du trapèze dans l'etude de la résistance des barrages en maçonnerie," *Compt. Rend. hebdom. Acad. Sci. Paris*, Vol. 126, pp. 1235–1240 (1898).
Fillunger, P., "Drei wichtige ebene spannungszustände des Keilförmigen körpers," *Zeit. Math. Physik.*, Vol. 60, pp. 275–285 (1912).
Carothers, S. D., "Plane strain in a wedge with applications to masonry dams," *Proc. Roy. Soc. Edinburgh*, Vol. 33, pp. 292–306 (1913).

APPENDIX A

The compatibility conditions

Consider the strain-displacement relations

$$\epsilon_{ij} = \frac{1}{2} (u_{i,j} + u_{j,i})$$ (A1)

where the comma denotes partial differentiation with respect to the coordinate variable x_i. If the displacement field u_i is given, then it is a relatively easy matter to compute the strain matrix ϵ_{ij}, by simply substituting the displacements into eq. (A1). The inverse problem of determining the displacement field from the strain field is not so straightforward. The difficulty arises from the fact that the three displacement functions u_i must be determined from the integration of the six first-order partial differential equations defined by (A1). Such a system is overdetermined, and for the existence of a single-valued and continuous solution for u_i (i.e., a deformed material region without cracks, discontinuities, material overlapping, etc.), it becomes necessary to impose certain restrictions on ϵ_{ij}. These conditions are referred to as the *compatibility conditions*. It must be noted that in a formulation in which the displacement components are chosen as the basic dependent variables, the compatibility equations are automatically satisfied, and no further consideration need be given to the compatibility conditions. If, on the other hand, the strain and stress components are chosen as the basic dependent variables without reference to the displacements, then the compatibility conditions are an essential part of the set of governing equations. In mathematical terms, the compatibility conditions are a statement of the conditions both *necessary* and *sufficient* for ensuring the single-valuedness of the displacement field u_i derived from the strain components ϵ_{ij}.

The necessary compatibility conditions can be obtained by eliminating the displacement components u_i from the strain-displacement relationships (A1). From the definition of the deformation gradient matrix $u_{i,j}$, we have

$$u_{i,j} = \epsilon_{ij} + \Omega_{ij}$$ (A2)

where Ω_{ij} is the rotation matrix. Differentiating (A2) we obtain

$$(u_{i,j})_{,k} = \epsilon_{ij,k} + \Omega_{ij,k}.$$ (A3)

Also from (A3) we have

$$(u_{i,k})_{,j} = \epsilon_{ik,j} + \Omega_{ik,j}.$$ (A4)

167

From (A3) and (A4) we obtain

$$\epsilon_{ij,k} - \epsilon_{ik,j} = \Omega_{ik,j} - \Omega_{ij,k}. \tag{A5}$$

From the definition of the rotation matrix we have

$$\Omega_{ik,j} - \Omega_{ij,k} = \Omega_{jk,i}. \tag{A6}$$

By substituting (A6) in (A5) and differentiating the result with respect to x_l we have

$$\epsilon_{ij,kl} - \epsilon_{ik,jl} = \Omega_{jk,il}. \tag{A7}$$

Similarly it can be shown that

$$\epsilon_{lj,ki} - \epsilon_{lk,ji} = \Omega_{jk,li}. \tag{A8}$$

Since the differentiations on the right-hand side of (A7) and (A8) commute, we have

$$\epsilon_{ij,kl} + \epsilon_{kl,ij} = \epsilon_{ik,jl} + \epsilon_{jl,ik}. \tag{A9}$$

These are the compatibility equations, which *must* be satisfied by the strain components if they are to be related to the displacements u_i through (A1), which in turn implies that u_i are single-valued and continuous. Consequently they are the *necessary* conditions.

Let us now focus attention on proving that the compatibility equations are also *sufficient* for generating a single-valued continuous displacement field u_i for simply connected domains. (A simply connected domain is one in which each and every closed curve can be continuously reduced to a point without passing out of the domain.) Consider a point $P(x_i)$ within the continuum at which the displacement field u_i^P and the rotation matrix Ω_{ij}^P are known. The displacement at any other location Q in the continuum can be represented by a line integral along a continuous curve from P to Q, i.e.

$$u_i^Q = u_i^P + \oint_P^Q du_i. \tag{A10}$$

We note that

$$du_i = \frac{\partial u_i}{\partial x_j} dx_j = u_{i,j} dx_j. \tag{A11}$$

Using (A2) and (A11) in (A10) we have

$$u_i^Q = u_i^P + \oint_P^Q \epsilon_{ij} dx_j + \oint_P^Q \Omega_{ij} dx_j. \tag{A12}$$

Since x_j^Q is a fixed location, we can replace dx_j by $d(x_j - x_j^Q)$. Hence, the last integral in (A12) can be integrated in parts as follows

$$\oint_P^Q \Omega_{ij} d(x_j - x_j^Q) = [\Omega_{ij}(x_j - x_j^Q)]_P^Q - \oint_P^Q (x_j - x_j^Q) d\Omega_{ij}. \tag{A13}$$

Noting that

$$d\Omega_{ij} = \Omega_{ij,k}dx_k \tag{A14}$$

the result (A13) can be rewritten as

$$\oint_P^Q \Omega_{ij}d(x_j - x_j^Q) = -\Omega_{ij}^P(x_j^P - x_j^Q) - \oint_P^Q (x_j - x_j^Q)\Omega_{ij,k}dx_k. \tag{A15}$$

From the definition of the rotation matrix Ω_{ij} we have

$$\Omega_{ij,k} = \frac{1}{2}[(u_{i,j})_{,k} - (u_{j,i})_{,k}]. \tag{A16}$$

By adding and subtracting the term $1/2u_{k,ij}$ to (A16) and rearranging terms, we have

$$\Omega_{ij,k} = \epsilon_{ik,j} - \epsilon_{jk,i}. \tag{A17}$$

Combining (A12), (A15), and (A17) we have

$$u_i^Q = u_i^P - \Omega_{ij}^P(x_j^P - x_j^Q) + \int_P^Q \Phi_{ik}dx_k \tag{A18}$$

where

$$\Phi_{ik} = \epsilon_{ik} - (x_j - x_j^Q)(\epsilon_{ik,j} - \epsilon_{jk,i}). \tag{A19}$$

For u_i^Q to be continuous and single-valued, the integral in (A18) must be path-independent. This implies that the integrand of (A18) must be an exact differential. From the theory of line integrals applicable to a simply connected domain, the necessary and sufficient condition for $\Phi_{ik}dx_k$ to be an exact differential is that

$$\Phi_{ik,l} = \Phi_{il,k}. \tag{A20}$$

From (A19) and (A20) we have

$$\epsilon_{ik,l} - x_{j,l}(\epsilon_{ik,j} - \epsilon_{jk,i}) - (x_j - x_j^Q)(\epsilon_{ik,jl} - \epsilon_{jk,il})$$
$$= \epsilon_{il,k} - x_{j,k}(\epsilon_{il,j} - \epsilon_{jl,i}) - (x_j - x_j^Q)(\epsilon_{il,jk} - \epsilon_{jl,ik}). \tag{A21}$$

Noting that $x_{i,j} = \delta_{ij}$, (A21) can be reduced to the condition

$$\epsilon_{ik,jl} + \epsilon_{jl,ik} = \epsilon_{il,jk} + \epsilon_{jk,il}. \tag{A22}$$

A simple interchange of indices will reveal that (A22) is identical to (A9). Consequently, we have proven that the conditions (A9) are also *sufficient* to ensure integrability of (A1) to generate a single-valued and continuous displacement field u_i. The mathematical aspects of the proof of the compatibility conditions applicable to continua were considered by a number of eminent elasticians including Saint-Venant (1864), Beltrami

(1886), Boussinesq (1871), and Cesaro (1906). A detailed exposition of the mathematical developments associated with the compatibility equations is given by Gurtin (1972).

Reference

Beltrami, E., "Osservazioni sulla nota precedente," *Atti Accad. Lincei Rend. Opere*, Vol. 4, pp. 510–512 (1892).

Boussinesq, J., "Étude nouvelle sur l'équilibre et le mouvement des corps solides élastiques dont certaines dimensions sont très petites par rapport à d'autres," *Premier Mémoire, J. Math. Pures Appl.*, Vol. 16, pp. 125–240 (1871).

Cesàro, E., "Sulle formole del Volterra, fondamentali nella teoria della distorsioni elastiche," *Rend. Napoli*, Vol. 12, pp. 311–324 (1906).

Saint Venant, Barré de A. J. C., Établissement élémentaire des formules et équations générales de la théorie de l'élasticité des corps solides," Appendix in: "Résumé des Leçons données à l'École des Ponts et Chaussées sur l'Application de la Mécanique," première partie, première section, *De la Résistance des Corps Solides*, par C.-L.M.H. Navier, 3rd Ed. Paris (1864).

Gurtin, M. E., "The linear theory of elasticity," In C. Truesdell (Ed.), *Handbuch der Physik*, Band IV a/2, Mechanics of Solids II, Springer-Verlag, Berlin (1972).

APPENDIX B

Cauchy's stress principle

The general theory of stress is due to Cauchy (1823). The stress principle can be stated as follows. Consider any closed surface ∂S within a continuum region B, which separates the region B into subregions B_1 and B_2. The interaction between these subregions can be represented by a field of stress vectors $T(\hat{n})$ defined on ∂S. By combining this principle with Euler's equations that express balance of linear momentum and moment of momentum in any kind of body, Cauchy derived a number of relationships that are now familiar, i.e.

$$T(\hat{n}) = -T(-\hat{n}) \tag{B1}$$

$$T(\hat{n}) = \sigma^T \hat{n} \tag{B2}$$

where $\hat{n}$ is the unit normal to ∂S and σ is the stress matrix. The second of these relationships is derived in Appendix C. Furthermore, in regions where the field variables have sufficiently smooth variations to allow spatial derivatives up to any order we have

$$\rho A = \text{div } \sigma + f \tag{B3}$$

where A is the acceleration field, and f is the body force per unit volume. This result expresses a necessary and sufficient condition for the balance of linear momentum, while, when (B3) is satisfied,

$$\sigma = \sigma^T \tag{B4}$$

is equivalent to the balance of moment of momentum with respect to an arbitrary point. In deriving (B4), it is implied that there are no body couples. If body couples and/or couple stresses [similar to those introduced by E. and F. Cosserat (1909)] are present, (B4) is modified but (B3) remains unchanged.

Cauchy's stress principle has four essential ingredients

(i) The physical dimensions of stress are (force)/(area).
(ii) Stress is defined on an imaginary surface that separates the region under consideration into two parts.
(iii) Stress is a vector or vector field equipollent to the action of one part of the material on the other.
(iv) The direction of the stress vector is not restricted.

171

References

Cauchy, A. L., "Recherches sur l'équilibre et le mouvement intérieur des corps solides ou fluides, élastiques ou non élastiques," *Bull. Soc. Philomath.*, Vol. 2, pp. 300–304 (1823).

Cosserat, E. and F., *Théorie des corps deformables*, Herrmann et Cie, Paris (1909).

APPENDIX C

Traction vector on an arbitrary plane

Consider the traction vectors T_i at the location x_i and referred to a plane the unit outward normal to which has components n_i. The state of stress at the same location is specified by σ_{ij}. From Cauchy's results we have

$$T_i = \sigma_{ji}n_j. \tag{C1}$$

The result (C1) can be proved by considering the Cauchy tetrahedron, which is a triangular pyramid bounded by parts of the three coordinate planes through 0 and a fourth plane ABC not passing through 0 (Figure C1). The tetrahedron could correspond to a location at either the boundary of a continuum region or at a surface within the continuum region.

The outward unit normal n_i is as shown in Figure C1. The direction cosines of the outward unit normal to ABC are

$$n_x = \cos(\angle AON); \; n_y = \cos(\angle BON); \; n_z = \cos(\angle CON). \tag{C2}$$

Considering the geometry of the tetrahedron, it can be shown that the height h of the tetrahedron is

$$h = (OA)n_x = (OB)n_y = (OC)n_z. \tag{C3}$$

Also the volume of the tetrahedral element ΔV is given by

$$\Delta V = \frac{1}{3}h\Delta S = \frac{1}{3}(OA)\Delta S_x = \frac{1}{3}(OB)\Delta S_y = \frac{1}{3}(OC)\Delta S_z \tag{C4}$$

where ΔS is the area of the plane ABC.
From (C3) and (C4), it is evident that

$$\Delta S_x = n_x\Delta S; \; \Delta S_y = n_y\Delta S; \; \Delta S_z = n_z\Delta S \tag{C5}$$

or in indicial notation

$$\Delta S_i = n_i\Delta S. \tag{C6}$$

Let us now consider the traction vectors acting on the tetrahedral element. The trac-

173

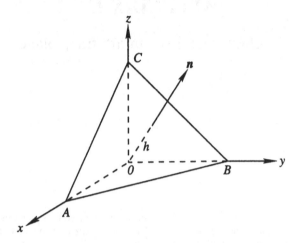

Figure C1   The Cauchy tetrahedron.

tion vectors on the plane surfaces through the three coordinate planes can be represented in terms of the stress components acting on these planes (Figure C2). The traction vector on the plane surface *ABC* has components T_x, T_y, and T_z. In addition to these traction vectors, it is assumed that the tetrahedron region is also subjected to body forces of intensity f_i having units of force per unit volume (Figure C3).

Consider the equilibrium of forces acting on the tetrahedral element. In the *x*-direction we have

$$T_x(\Delta S) - \sigma_{xx}(\Delta S_x) - \sigma_{yx}(\Delta S_y) - \sigma_{zx}(\Delta S_z) - \Delta V f_x = 0. \qquad (C7)$$

Using (C4) and (C5) in (C7) we have

$$T_x - \sigma_{xx}n_x - \sigma_{yx}n_y - \sigma_{zx}n_z - \frac{1}{3}hf_x = 0. \qquad (C8)$$

We now let *h* approach zero. In doing so, we allow the arbitrary plane on which the tractions T_i are prescribed to also coincide with the location at which the stress matrix σ_{ij} is prescribed. In the limit as $h \to 0$, (C8) reduces to

$$T_x = \sigma_{xx}n_x + \sigma_{yx}n_y + \sigma_{zx}n_z = \sigma_{jx}n_j. \qquad (C9)$$

Similarly, by considering the equilibrium of the tetrahedral element in the *y*- and *z*-directions we can show that

$$T_y = \sigma_{jy}n_j; \; T_z = \sigma_{jz}n_j. \qquad (C10)$$

Combining (C9) and (C10) we can write the final result in indicial form as follows

$$T_i = \sigma_{ji}n_j. \qquad (C11)$$

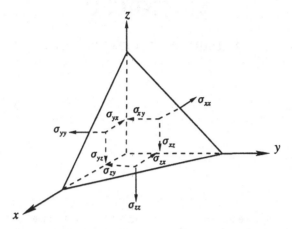

Figure C2 Stress components on the faces of the tetrahedron perpendicular to the co-ordinate direction.

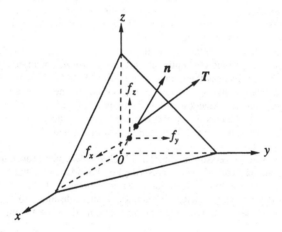

Figure C3 The traction vector T acting on the oblique face of the tetrahedron and the components of the body force vector f.

The result (C11) confirms Cauchy's result that allows us to determine the tractions at a point acting on an arbitrary plane through the point when the tractions on three mutually perpendicular planes through the point are known.

APPENDIX D

Uniqueness of solutions in classical elasticity theory

Let us suppose that we have been able to obtain a solution to a particular problem in linear elasticity theory, which satisfies all the governing equations (i.e., equations of equilibrium, compatibility, and stress-strain relations) and the boundary conditions that include both displacement and traction boundary conditions are specified, respectively, on surfaces S_D and S_T of the boundary S ($= S_D \cup S_T$). The natural question that can be asked is whether the stresses and strains in the medium can be obtained without ambiguity. There are a number of ways in which this question can be answered. These range from the rigorous mathematical study of existence and uniqueness applicable to some fundamental differential equation governing the elasticity problem (e.g., the biharmonic equation) to the conventional *reductio ad absurdum* approach outlined in this section. The uniqueness theorem is due to Kirchhoff (1850), and it proves that if either the displacements or tractions are specified on the boundaries of a three-dimensional body, then uniqueness of the elasticity solution is assured, and there is at most one solution in the classical sense, provided there is energy stored in the elastic body, as strain energy, during the deformation, and that λ and G obey the inequalities $(3\lambda + 2G) > 0$ and $G > 0$.

We shall consider an elastic medium in which displacement boundary conditions and traction boundary conditions are prescribed on boundaries S_D and S_T as shown in Figure D1. As a prelude to the proof, let us remind ourselves that for an elastic body, there exists a strain energy function W that is a positive definite, homogeneous quadratic function of the strains with appropriately symmetrical coefficients such that

$$\sigma_{ij} = \frac{\partial W}{\partial \epsilon_{ij}} \tag{D1}$$

and

$$W = \frac{1}{2}\,\sigma_{ij}\epsilon_{ij}. \tag{D2}$$

For an isotropic elastic body, B, the strain energy function can be further expressed as

$$W = \frac{1}{2}\,\lambda(\epsilon_{kk})^2 + G\epsilon_{ij}\epsilon_{ij}. \tag{D3}$$

176

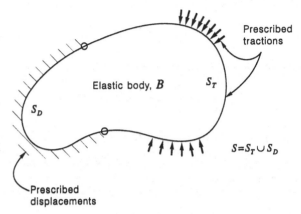

Figure D1 Elastic body in equilibrium under specified boundary conditions.

We have shown in Chapter 2 that W is positive definite provided $E > 0$ and $-1 < \nu \le 1/2$.

Let us now concentrate on proving the classical uniqueness theorem due to Kirchhoff. To start with, assume that there are two sets of solutions with stress, strain, and displacement fields defined by $\sigma_{ij}^{(1)}$, $\epsilon_{ij}^{(1)}$, $u_i^{(1)}$ and $\sigma_{ij}^{(2)}$, $\epsilon_{ij}^{(2)}$, and $u_i^{(2)}$. In general these fields will both satisfy the field equations

$$\sigma_{ij,j} + f_i = 0 \quad \text{for} \quad x_i \epsilon B \tag{D4}$$

$$T_i = T_i^* \quad \text{for} \quad x_i \epsilon S_T \tag{D5}$$

$$u_i = u_i^* \quad \text{for} \quad x_i \epsilon S_D \tag{D6}$$

where T_i^* and u_i^* are prescribed.

If the two sets are solutions to the same problem, then $\tilde{\sigma}_{ij} = \sigma_{ij}^{(2)} - \sigma_{ij}^{(1)}$; and $\tilde{\epsilon}_{ij} = \epsilon_{ij}^{(2)} - \epsilon_{ij}^{(1)}$ would be a solution to the problem where $f_i \equiv 0$ at every point in B; and $\tilde{T}_i \tilde{u}_i \equiv 0$ at every point of the boundary S, since on the boundary either $\tilde{T}_i = 0$ or $\tilde{u}_i = 0$. Integrating over the entire boundary $S = S_T \cup S_D$ we should have

$$\int_S \tilde{T}_i \tilde{u}_i dS = 0. \tag{D7}$$

Considering boundary equilibrium of external tractions and internal stresses (see, e.g., (1.12) or Appendix C) we have

$$\tilde{T}_i = \tilde{\sigma}_{ij} n_j. \tag{D8}$$

Then, substituting (D8) in (D7) we have

$$\int_S \tilde{\sigma}_{ij} \tilde{u}_i n_j dS = 0. \tag{D9}$$

From Green's theorem in integral calculus we have

$$\int_S \tilde{\sigma}_{ij}\tilde{u}_i n_j dS = \int_B (\tilde{\sigma}_{ij}\tilde{u}_i),_j dV = 0. \tag{D10}$$

Consider the integrand of the second equation of (D10)

$$\frac{\partial}{\partial x_j}(\tilde{\sigma}_{ij}\tilde{u}_i) = \tilde{\sigma}_{ij}\frac{\partial \tilde{u}_i}{\partial x_j} + \tilde{u}_i\frac{\partial \tilde{\sigma}_{ij}}{\partial x_j}. \tag{D11}$$

We note that, by virtue of the compatibility equations, the displacement field is single-valued in any simply connected region (see, e.g., Appendix A); as such, derivatives of $\tilde{u}_i$ can be obtained up to any required order. Since $\tilde{\sigma}_{ij}$ satisfies the equations of equilibrium, in the absence of body forces, the second term on the right-hand side of (D11) is zero. Also, from results given in Chapter 1 (eqs. 1.3 and 1.4), we have

$$\frac{\partial \tilde{u}_i}{\partial x_j} = \tilde{\epsilon}_{ij} + \tilde{\Omega}_{ij} \tag{D12}$$

where $\tilde{\epsilon}_{ij}$ and $\tilde{\Omega}_{ij}$ are, respectively, the strain and rotation tensors. Consequently, we have

$$\tilde{\sigma}_{ij}\frac{\partial \tilde{u}_i}{\partial x_j} = \tilde{\sigma}_{ij}\tilde{\epsilon}_{ij} + \tilde{\sigma}_{ij}\tilde{\Omega}_{ij}. \tag{D13}$$

Since $\tilde{\sigma}_{ij} = \tilde{\sigma}_{ji}$ and $\tilde{\Omega}_{ij} = -\tilde{\Omega}_{ji}$, the second term in the right-hand side of (D13) is identically equal to zero. Hence, using (D7), (D10), (D11), and (D13) we have, with (D2)

$$\int_S \tilde{T}_i\tilde{u}_i dS = \int_B \tilde{\sigma}_{ij}\tilde{\epsilon}_{ij}dV = \int_B 2W dV = 0. \tag{D14}$$

Since W is positive definite and continuous, the last integral in (D14) can vanish only if $W \equiv 0$ everywhere in the domain. This is possible if, and only if, $\tilde{\epsilon}_{ij} = 0$; consequently

$$\epsilon_{ij}^{(1)} \equiv \epsilon_{ij}^{(2)} \tag{D15}$$

and by virtue of Hooke's law

$$\sigma_{ij}^{(1)} \equiv \sigma_{ij}^{(2)}. \tag{D16}$$

These arguments explain very simply that it is not possible for two different states of stress and strain to exist in a region of an elastic continuum for a given set of body forces, displacement boundary conditions, and traction boundary conditions.

In deriving this classical proof of uniqueness, it is implicitly assumed that the strain fields are small and that the stresses associated with the displacement and traction boundary conditions and body forces are described by the Cauchy stress components referred to the undeformed body. Consequently, the uniqueness theorem will not apply to situations involving large deflections or large strains. Examples of such deformations include buckling of elastic bodies or problems related to elastic stability. Extensive historical reviews and mathematically oriented accounts of the subject of uniqueness of solutions in classical elasticity theory are given in the review articles and texts by Sternberg (1960), Knops and Payne (1970, 1971), and Gurtin (1972).

References

Gurtin, M. E., "The linear theory of elasticity," In C. Truesdell (Ed.), *Handbuch der Physik*, Band IV a/2, Mechanics of Solids II, Springer-Verlag, Berlin (1972).

Kirchhoff, G., "Über das Gleichgewicht und die Bewegung einer elastichen Scheibe," *J. Reine Angew. Math.*, Vol. 40, pp. 51–88 (1850).

Knops, R. J., and Payne, L. E., "On uniqueness and continuous dependence in dynamical problems of thermoelasticity," *Int. Jour. Solids Structs.*, Vol. 6, pp. 1173–1184 (1970).

Knops, R. J., and Payne, L. E., *Uniqueness Theorems in Linear Elasticity*, Springer Tracts in Natural Philosophy, Vol. 19, Springer Verlag, Berlin (1971).

Sternberg, E., "On some recent developments in the linear theory of elasticity," In J. N. Goodier and N. J. Hoff (Eds.), *Structural Mechanics*, Pergamon, New York (1960).

APPENDIX E

Saint-Venant's principle

In 1855, Barre de Saint-Venant postulated a principle that conforms well with the application of the classical theory of elasticity to problems of engineering interest. The principle was enunciated primarily in connection with Saint-Venant's celebrated studies related to the extension, flexure, and torsion of prismatic and cylindrical bodies. The first universal statement of the principle is attributed to Boussinesq (1885) and reads as follows:

An equilibrated system of external forces applied to an elastic body, all of the points of application lying within a given sphere, produces deformations of negligible magnitude at distances from the sphere which are sufficiently large compared to its radius.

Love (1927) interprets this as follows: "According to this principle, the strains that are produced in a body by the application, to a small part of its surface, of a system of forces statically equivalent to zero force and zero couple, are of negligible magnitude at distances which are large compared with the linear dimensions of this part." As pointed out by von Mises (1945), these statements are in need of clarification since the forces applied to a body at rest must be in equilibrium in any event. Only when the body is of infinite extent and provided we require the tractions at infinity to vanish suitably, is it meaningful to speak of a nonequilibrated system of forces applied to a part of its bounding surface? In this particular case, the strains produced by a given surface-loading are sufficiently small at locations far from the loaded areas regardless of whether or not the applied forces are self-equilibrated. In addition, the stresses and strains at a fixed point of an elastic body, in the absence of body forces, may be made arbitrarily large or small by suitably choosing the magnitude of the applied loads. Despite these reservations, the importance of Saint-Venant's principle can hardly be overstated since there are only a few physical problems to which the theory of elasticity can be applied without invoking it. The principle has acquired, in structural mechanics and stress analysis, not only a prominent role but also indiscriminate use (Selvadurai, 1979), as it provides a justification for ignoring major difficulties in a large number of problems. Since its inception, numerous researchers have attempted to justify the principle or point out its limitations. Reasonably comprehensive bibliographies of works that pertain to Saint-Venant's principle are given by Gurtin (1972).

A statement of Saint-Venant's principle that is admittedly not complete or mathematically rigorous but which is useful and excludes against possible exceptions can be as follows (Selvadurai, 1979):

Consider a distribution of forces on a small region of a homogeneous isotropic elastic body $\partial \Sigma$. These forces are resisted by primarily the material in the neighborhood of $\partial \Sigma$. The stresses resulting from this applied loading will fall off rapidly away from the loaded region $\partial\Sigma$ and become negligible at distances large compared with the dimensions of $\partial\Sigma$. Let us now assume that the forces acting on $\partial\Sigma$ are replaced by a "statically equivalent" distribution over the same region $\partial\Sigma$. The term statically equivalent implies that the two distributions of forces have the same resultant force and moment. Saint-Venant's principle assures us that the effects of the two different distributions of forces on $\partial\Sigma$, at distances large in comparison to the dimension of $\partial\Sigma$, are essentially the same.

Let us now try to investigate the validity of the principle by appeal to a geomechanics problem, particularly in relation to the loading of an elastic half-space. Consider Boussinesq's classical problem of the loading of an isotropic elastic half-space by a force P acting normal to its surface (Figure E1(a)). As we have discussed in Chapter 3, there is an exact solution to this problem that

(i) satisfies all the necessary traction boundary conditions on the surface of the half-space region,
(ii) gives rise to stress and displacement fields that decay to zero as both r and z tend to infinity, and
(iii) have force resultants on either any hemispherical surface centered at the origin or any plane $z =$ constant, equivalent to an axial force P.

To apply Saint-Venant's principle to this problem, we need to represent the concentrated force P by statically equivalent distributions acting in the neighborhood of P. Now, the term "neighborhood" is difficult to interpret in the classical Boussinesq's problem since the point force has no natural length parameter associated with it. Let us rectify this ambiguity by considering statically equivalent distributions, which can be assigned a length parameter. How many such statically equivalent distributions can we consider? There are an infinite number of possibilities. To keep the analysis to a manageable level let us consider the following four different representations of the concentrated force P. The statical equivalence between any assumed distributions and the axisymmetric concentrated load implies that the assumed distributions must necessarily be axisymmetric. Bearing this in mind, we choose the following:

(i) a ring load of radius a and intensity $P/2\pi a$ (Figure E1(b)),
(ii) a conical circular load of radius a and peak stress intensity $3P/\pi a^2$ (Figure E1(c)),
(iii) a uniform circular load of radius a and stress intensity $P/\pi a^2$ (Figure E1(d)), and
(iv) a rigid circular load of radius a and stress intensity $P/2\pi a(a^2 - r^2)^{1/2}$ (Figure E1(e)).

It can be verified quite easily that all these distributions contribute to a total axisymmetric load of magnitude P. Now, the rest remains for us to verify that all these four distributions of loading on the surface of the half-space region will give rise to displacement and stress fields that converge to approximately the same result as $r, z \to \infty$. Admittedly, if we are to rigorously examine the decay processes in the displacements and the stresses, we should consider all the displacement components (u_r, u_z) and all the stress components (σ_{rr}, $\sigma_{\theta\theta}$, σ_{zz}, σ_{rz}) associated with the axisymmetric loading configurations. In the interests of keeping the discussion to illustrating the basic process of decay in the effects of the loading, let us restrict attention to the evaluation of the displacement u_z along the z-axis.

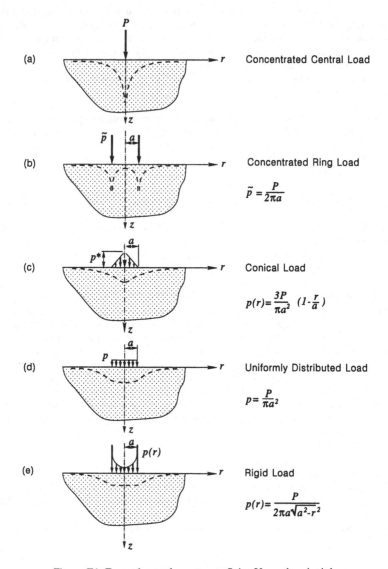

Figure E1   Examples to demonstrate Saint-Venant's principle.

(i) Concentrated loading

From Boussinesq's solution for u_z we have

$$u_z(r,z) = \frac{P(1 - \nu)}{4\pi G}\left[\frac{2}{(r^2 + z^2)^{1/2}} + \frac{z^2}{(1 - \nu)(r^2 + z^2)^{3/2}}\right]. \qquad (E1)$$

Let us now introduce a length parameter a and represent (E1) in the form

$$u_z(r,z) = \frac{P(1 - \nu)}{4\pi Ga} \left[\frac{2a}{(r^2 + z^2)^{1/2}} + \frac{z^2 a}{(1 - \nu)(r^2 + z^2)^{3/2}} \right]. \quad \text{(E2)}$$

Also

$$[u_z(0,z)]_{\text{concentrated load}} = \frac{P(1 - \nu)}{4\pi Ga} \left[\frac{(3 - 2\nu)}{(1 - \nu)} \left(\frac{a}{z} \right) \right]. \quad \text{(E3)}$$

As is evident $\{u_z(0,z)\}_{\text{concentrated load}} \to 0$ as $z \to \infty$.

(ii) Ring load

The displacement along the z-axis due to the ring load can be obtained by integrating Boussinesq's result for the concentrated load. For an element of the ring load of length $ad\theta$, the incremental displacement along the axis is

$$du_z(0,z) = \left(\frac{P}{2\pi a} \, ad\theta \right) \left(\frac{(1 - \nu)}{4\pi G} \right) \left[\frac{2}{(a^2 + z^2)^{1/2}} + \frac{z^2}{(1 - \nu)(a^2 + z^2)^{3/2}} \right]. \quad \text{(E4)}$$

Hence, the complete displacement due to the ring load is given by integrating (E4) for $\theta \epsilon (0, 2\pi)$, i.e.

$$[u_z(0,z)]_{\text{ring load}} = \frac{P(1 - \nu)}{2\pi Ga} \left[\frac{2a}{(a^2 + z^2)^{1/2}} + \frac{z^2 a}{(1 - \nu)(a^2 + z^2)^{3/2}} \right]. \quad \text{(E5)}$$

Let us now expand the terms of (E5) within the square brackets in powers of $(1/z)$; i.e.

$$\frac{a}{(a^2 + z^2)^{1/2}} = \left(\frac{a}{z} - \frac{1}{2} \frac{a^3}{z^3} + \ldots \right) \quad \text{(E6)}$$

$$\frac{z^2 a}{(a^2 + z^2)^{3/2}} = \left(\frac{a}{z} - \frac{3}{2} \frac{a^3}{z^3} + \ldots \right). \quad \text{(E7)}$$

Hence the dominant variation of (E6) and (E7) as $z \to \infty$ corresponds to (a/z). Substituting the dominant terms of these variations in (E5) we have

$$[u_z(0,z)]_{\text{ring load}} = \frac{P(1 - \nu)}{4\pi Ga} \left[\frac{(3 - 2\nu)}{(1 - \nu)} \frac{a}{z} \right] \quad \text{(E8)}$$

which is identical to that obtained for the concentrated load P.

(iii) Conical circular load

The displacement along the axis due to the conical circular distributed loading can be obtained by integrating Boussinesq's solution for the concentrated load. The distribution of the conical load is (see Figure E1(c))

$$\sigma_{zz}(r,0) = \frac{3P}{\pi a^2}\left(1 - \frac{r}{a}\right).$$ (E9)

For an element of the conical load of area $r\,dr\,d\theta$, the incremental displacement along the z-axis is

$$du_z(r,0) = \frac{3P}{\pi a^2}\left(1 - \frac{r}{a}\right)r\,dr\,d\theta\left(\frac{(1-\nu)}{4\pi G}\right)\left[\frac{2}{(r^2+z^2)^{1/2}}\right.$$

$$\left. + \frac{z^2}{(1-\nu)(r^2+z^2)^{3/2}}\right].$$ (E10)

The complete displacement distribution along the z-axis is given by

$$[u_z(0,z)]_{\text{conical load}} = \frac{6P(1-\nu)}{4\pi G a^2}\int_0^a\left(r - \frac{r^2}{a}\right)\left[\frac{2}{(r^2+z^2)^{1/2}} + \right.$$

$$\left. \frac{z^2}{(1-\nu)(r^2+z^2)^{3/2}}\right]dr.$$ (E11)

Performing the integrations in (E11) we have

$$[u_z(0,z)]_{\text{conical load}} = \frac{6P(1-\nu)}{4\pi G a}\left[\frac{(a^2+z^2)^{1/2}}{a} - \frac{z}{a}\left(\frac{1-2\nu}{1-\nu}\right)\right.$$

$$\left. - \frac{\nu}{(1-\nu)}\frac{z^2}{a^2}\ell_n\left\{\frac{a+\sqrt{a^2+z^2}}{a}\right\}\right].$$ (E12)

Let us now examine the dominant behavior of (E12) as (z/a) becomes large; to do this, expand $(a^2+z^2)^{1/2}$ and the logarithmic term in powers of (a/z). We can write

$$\frac{z}{a}\left(1 + \frac{a^2}{z^2}\right)^{1/2} = \frac{z}{a} + \frac{1}{2}\frac{a}{z} + \dots$$ (E13)

$$\ell_n\left\{\frac{a+\sqrt{a^2+z^2}}{z}\right\} = \frac{a}{z} - \frac{1}{6}\frac{a^2}{z^2} + \dots.$$ (E14)

Substituting (E13) and (E14) in (E12) we obtain the far-field dominant behavior as

$$[u_z(0,z)]_{\text{conical load}} = \frac{P(1-\nu)}{4\pi G a}\left[\frac{(3-2\nu)}{(1-\nu)}\frac{a}{z}\right].$$ (E15)

(iv) Uniform circular load

Here, the analysis is straightforward, and we shall simply quote the final result for the distribution of displacement along the axis of the uniform circular load (Figure E1(d)).

$$[u_z(0,z)]_{\text{uniform circular load}} = \frac{P(1-\nu)}{4\pi Ga}\left[4\left\{\frac{\sqrt{a^2+z^2}}{a} - \frac{z}{a}\right\}\right.$$
$$\left. \times \left\{1 + \frac{z}{2(1-\nu)(a^2+z^2)^{1/2}}\right\}\right] \qquad (E16)$$

Again, expanding the term within the right-hand-side square bracket, we can show that the dominant decay behavior of the axial displacement due to the uniform circular load is

$$[u_z(0,z)]_{\text{uniform circular load}} = \frac{P(1-\nu)}{4\pi Ga}\left[\frac{(3-2\nu)}{(1-\nu)}\left(\frac{a}{z}\right)\right]. \qquad (E17)$$

(v) Rigid load

Up to now, we have considered situations in which the tractions were prescribed within the region $0 < r < a$. Let us now consider the situation where uniform displacement is prescribed over the region $0 \le r \le a$. We have discussed this problem in Chapter 4. The contact stress distribution over the region $0 \le r \le a$, which is equivalent to an axial force P, is given by

$$\sigma_{zz}(r,0) = \frac{P}{2\pi a}\frac{1}{\sqrt{a^2-r^2}}. \qquad (E18)$$

For an element of this contact stress distribution, the increment of displacement along the z-axis is given by

$$du_z(0,z) = \frac{P(1-\nu)rdrd\theta}{2\pi a\sqrt{a^2-r^2}(4\pi G)}\left[\frac{2}{(r^2+z^2)^{1/2}} + \frac{z^2}{(1-\nu)(r^2+z^2)^{3/2}}\right]. \qquad (E19)$$

Integrating (E19) over the region of application of the rigid load we have

$$[u_z(0,z)]_{\text{rigid load}} = \frac{P(1-\nu)}{4\pi Ga}\left[2\tan^{-1}\left(\frac{a}{z}\right) + \frac{az}{(1-\nu)(a^2+z^2)}\right]. \qquad (E20)$$

It is easy to verify that when $z = 0$

$$[u_z(0,0)]_{\text{rigid load}} = \frac{P(1-\nu)}{4Ga} \qquad (E21)$$

which is Boussinesq's classical result for the displacement of a smooth rigid circular foundation of radius a resting on the surface of an isotropic elastic half-space.

In order to examine the behavior of the solution as (z/a) becomes large, let us once again expand the terms within the right-hand-side square bracket of (E20) as power series in (a/z); we have

$$\tan^{-1}\left(\frac{a}{z}\right) = \frac{a}{z} - \frac{a^3}{3z^3} + \dots \tag{E22}$$

$$\frac{1}{[1 + (a/z)^2]} = 1 - \frac{a^2}{z^2} + \dots \tag{E23}$$

Substituting the power series (E22) and (E23) in (E20) we have

$$[u_z(0,z)]_{\text{rigid load}} = \frac{P(1 - \nu)}{4\pi Ga}\left[\frac{(3 - 2\nu)}{(1 - \nu)}\left(\frac{a}{z}\right)\right]. \tag{E24}$$

If we think a little bit about what we have done, several things should come to mind. Firstly, all equivalent load representations of the concentrated force give rise to exactly the same asymptotic or far-field behavior in evaluating the displacement along the axis of symmetry. We have performed the calculations only for the displacement field along the z-axis. It is also possible to evaluate other salient displacements, such as $u_z(r,0)$, $u_r(r,0)$, etc., and, for that matter, the stress components σ_{ij}, and conclusively prove that the far-field or the asymptotic behavior as $[(r/a)^2 + (z/a)^2]^{1/2} \to \infty$ is invariant of the precise nature of the axisymmetric load distribution applied at the surface of the half-space region. Secondly, we have used very elementary mathematical manipulations to prove, at least in a specialized sense, the most formidable principle of Saint-Venant. (It must indeed be gratifying to most engineering students to observe that the mathematics they have learned in the early years of an undergraduate program can be put to good use!)

The methodology adopted in illustrating the validity of Saint-Venant's principle to isotropic elastic bodies is more heuristic than rigorous. That is, we have learned from experiences of a number of examples that the principle does in fact work. There are, of course, more sophisticated formal proofs of the principle, attributed to a number of eminent investigators including Goodier, von Mises, Sternberg, Naghdi, Toupin, Hoff, Horvay, and Horgan. The complete references to these mathematical expositions of the principle, which range from the examination of the asymptotic solutions, to the class of differential equations governing elasticity problems, to consideration of the decay of elastic energy, can be found in the elegant review article by Gurtin (1972).

An obvious question that would come to mind would be "Are there any situations in elasticity in which Saint-Venant's principle will not be satisfied?" There are instances in which the effects of applied load can be transmitted to significant distances in comparison with the dimensions of the loaded region. A notable example is the case of highly anisotropic elastic materials or elastic materials that are reinforced with inextensible fibers. In this case, the stresses and displacements are channeled in the directions of the higher stiffnesses. Examples of such phenomena are documented by Horgan (1972) and Spencer (1972). For the purposes of geomechanical applications, the range of geomaterial anisotropies encountered in practice are not sufficiently large to warrant negation of Saint-Venant's principle. This principle offers a powerful and convenient method for evaluating the influences of loads applied to elastic geomaterials and the manner in which the effects of these loads are realized at remote locations, particularly by appeal to very simple analytical results.

References

Gurtin, M. E., "The linear theory of elasticity," In C. Truesdell (Ed.), *Handbuch der Physik*, Band IV a/2, Mechanics of Solids II, Springer-Verlag, Berlin (1972).

Horgan. C. O., "On Saint-Venant's principle in plane anisotropic elasticity," *Jour. Elasticity*, Vol. 2, pp. 169–180 (1972).

Love, A. E. H., *A Treatise on the Mathematical Theory of Elasticity*, Cambridge University Press, London (1927).

Saint-Venant, Barré de A. J. C., "Mémoire sur la torsion des prismes, avec des considérations sur leur flexion, ainsi que sur l'équilibre intérieur des solides élastiques en général et des formules pratiques pour le calcul de leur résistance à divers efforts s'exerçant simultanément," *Mem. Savants Etrangers*, T 14, pp. 233–560 (1855).

Selvadurai, A. P. S., *Elastic Analysis of Soil-Foundation Interaction*, Elsevier, Amsterdam (1979).

Spencer, A. J. M., "Deformations of fibre reinforced materials," *Oxford Science Research Papers*, Clarendon Press, Oxford (1972).

von Mises, R., "On Saint-Venant's principle," *Bull. Amer. Math. Soc.*, Vol. 51, pp. 555–562 (1945).

APPENDIX F

Principles of virtual work

The principles of virtual work are useful approaches to the solution of many problems in structural mechanics and applied mechanics. The principles are applicable to problems dealing with materials that exhibit both elastic and inelastic constitutive behavior and to problems that deal with dynamic and stability effects. The principles of virtual work are presented here in order to facilitate their application to the derivation in Appendix G of reciprocal theorems applicable to elastic bodies.

Consider an elastic medium restrained against rigid motion and subjected to a set of forces M_i. The state of stress in the medium at an arbitrary point P due to M_i is denoted by σ_{ij}, and the corresponding displacement and strain fields are denoted by u_i and ϵ_{ij}, respectively. Since the body is in equilibrium we require

$$\sigma_{ij,j} + f_i = 0 \tag{F1}$$

where f_i are the body forces. Now consider the case where the deformed body is subjected to a second system of forces (Figure F1), which produces a kinematically admissible set of displacements. An imaginary infinitesimal deformation of this type is called a *virtual deformation* and the associated displacements are referred to as *virtual displacements*. We denote the virtual displacement at point 'i' in the direction of M_i by δm_i. Similarly, the virtual displacement at any generic point P is denoted by δu_i, and the corresponding virtual strains are

$$\delta \epsilon_{ij} = \frac{1}{2} (\delta u_{i,j} + \delta u_{j,i}). \tag{F2}$$

The external virtual work (i.e., the work of the real loads M_i moving through the virtual displacements δm_i) is denoted by δW_E. It is possible to show that this virtual work is equal to the change in strain energy δU resulting from the real stresses σ_{ij} undergoing virtual strains $\delta \epsilon_{ij}$. To demonstrate this, consider the virtual work of the tractions T_i acting on the surface of the structure and the body forces f_i acting within the elastic body, given by

$$\delta W_E = \int_S T_i(\delta u_i)dS + \int_V f_i(\delta u_i)dV \tag{F3}$$

188

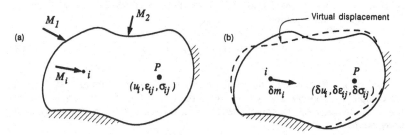

Figure F1 Loading of a continuum region by generalized forces M_i and by virtual displacements δu_i.

since $T_i = \sigma_{ij} n_j$, the surface integral in (F3) can be written as

$$\int_S T_i(\delta u_i)dS + \int_S \sigma_{ij} n_j(\delta u_i)dS. \tag{F4}$$

From Green's theorem in integral calculus we have (see e.g., (D10))

$$\int_S \sigma_{ij}(\delta u_i)n_j dS = \int_V \{\sigma_{ij}(\delta u_i)\}_{,j} dV. \tag{F5}$$

Hence

$$\delta W_E = \int_V [\{\sigma_{ij}(\delta u_i)\}_{,j} + f_i(\delta u_i)]dV. \tag{F6}$$

Now, since σ_{ij} is symmetric and the body is in equilibrium, (Eq. F6) reduces to

$$\delta W_E = \int_V \sigma_{ij}(\delta \epsilon_{ij})dV = \delta U \tag{F7}$$

where δU is the change in strain energy resulting from the stresses σ_{ij} undergoing virtual strains $\delta \epsilon_{ij}$.

In the case where the elastic body is subjected to a system of discrete external forces V_i and zero body forces, the result for the external virtual work δW_E can be expressed as a summation of discrete quantities $V_i \delta v_i$, where δv_i are the virtual displacements in the direction of V_i; i.e.

$$\delta W_E = \sum_{i=1}^{n} V_i(\delta v_i). \tag{F8}$$

This completes the development of the concept of virtual work. The same procedure can be used to develop an analogous relationship governing *complementary virtual work*. In this instance, a set of virtual surface tractions δT_i and virtual body forces

δf_i result in a state of virtual stress $\delta \sigma_{ij}$. These fictitious forces and stresses must satisfy the equilibrium equations

$$(\delta \sigma_{ij})_{,j} + (\delta f_i) = 0. \tag{F9}$$

Except for this constraint, the virtual tractions δT_i and the virtual body forces δf_i are completely arbitrary. The work done by the virtual forces on the actual deformation u_i is referred to as the complementary virtual work δW_E^*. Considering the elastic body, we have

$$\delta W_E^* = \int_S (\delta T_i) u_i dS + \int_V (\delta f_i) u_i dV. \tag{F10}$$

Using procedures similar to those outlined previously we can show that

$$\delta W_E^* = \int_V (\delta \sigma_{ij}) \epsilon_{ij} dV = \delta U^* \tag{F11}$$

where δU^* is the complementary strain energy. Again, if the virtual forces δV_i^* are applied at discrete locations and the actual displacements are v_i^*, the expression for the external complementary virtual work in this case is

$$\delta W_E^* = \sum_{i=1}^{n} (\delta V_i^*) v_i^*. \tag{F12}$$

APPENDIX G

Betti's reciprocal theorem

In 1872, Enrico Betti, an Italian mathematician, put forward a reciprocal theorem that is one of the most significant results in the classical theory of elasticity. The generalized proof of the theorem can be approached in the following way. Consider a linearly elastic body B, loaded by a set of forces F_i ($i = 1, 2, \ldots, n$). The displacements in the direction of F_i are denoted by w_i and displacements at an arbitrary location due to F_i are denoted by w_j. The elastic body is now subjected to forces F_j^* ($j = 1, 2, \ldots, m$) in the direction of w_j. The displacements in the direction of F_j^* are denoted by w_j^*, and the displacements in the direction of F_i are denoted by w_i^* (Figure G1). The displacement, strain, and stress fields in the two states are denoted by (u_i, ϵ_{ij}, σ_{ij}) and (u_i^*, ϵ_{ij}^*, σ_{ij}^*), respectively. Consider the two loading scenarios where, in the first, F_i is applied first followed by F_j^*, and, in the second, F_j^* is applied first followed by F_i.

If we consider the forces F_i as the real loads and F_j^* as the force system that produces the virtual displacements w_i^* and w_j^*, the principle of virtual work (Appendix F) gives

$$\sum_{i=1}^{n} F_i w_i^* = \int_V \sigma_{ij} \, \epsilon_{ij}^* \, dV. \tag{G1}$$

Consider now the case where F_j^* are the real loads and F_i is the force system that produces the virtual displacements w_i and w_j. Hence, from the principle of virtual work,

$$\sum_{j=1}^{m} F_j^* \, w_j = \int_V \sigma_{ij}^* \, \epsilon_{ij} \, dV. \tag{G2}$$

From the equations of elasticity we have

$$\sigma_{ij} = E_{ijkl} \, \epsilon_{kl}$$

$$\sigma_{ij}^* = E_{ijkl} \, \epsilon_{kl}^* \tag{G3}$$

where E_{ijkl} is the elasticity matrix. Using these relationships in (G1) and (G2) we have

$$\sum_{i=1}^{n} F_i w_i^* = \int_V E_{ijkl} \, \epsilon_{kl} \, \epsilon_{ij}^* \, dV \tag{G4}$$

191

Figure G1 Reciprocal states.

and

$$\sum_{i=1}^{m} F_j^* \, w_j = \int_V E_{ijkl} \, \epsilon_{ij} \, \epsilon_{kl}^* \, dV. \tag{G5}$$

If the elasticity matrix is symmetric

$$E_{ijkl} = E_{klij} \tag{G6}$$

then by interchanging the indices in the right-hand side of (G5), it is evident that (G5) is identical to (G4). Consequently we obtain the result

$$\sum_{i=1}^{n} F_i \, w_i^* = \sum_{j=1}^{m} F_j^* \, w_j. \tag{G7}$$

As a particular case, we can consider the loading systems to correspond to single *unit forces* applied at the respective locations. Then (G7) reduces to

$$w_i^* = w_j \tag{G8}$$

which is the law of reciprocal displacements attributed to Maxwell (1864). Maxwell developed this principle during the course of investigations relating to the behavior of elastic structural frames. The law states that the displacement at i due to a unit force at j is equal to the displacement at j due to the unit force at i. The applicability of the law extends to "generalized unit forces." Hence the "unit force" at i could be a "force," and the "unit force" at j could be a moment. Consequently, w_i^* is a displacement, but w_i is a rotation (Figure G2).

The importance of these theorems to linear elasticity is quite evident. Firstly, for the validity of the principle, the elasticity matrix must be symmetric. Or conversely, if reciprocity exists, the elasticity matrix must be symmetric. This aspect becomes particularly important in materials testing for the determination of the elastic constants. Secondly, the facility to relate two states of loading through their corresponding displacement fields can be put to considerable advantage in examining problems where, for example, the solution for an auxiliary problem required for the application of the reciprocal theorem has a particularly simple form. The proof presented in the preceding section is a relatively simple version that does not place any restriction on the shape of the domain B, material variability within B, etc. A complete account of the proof is given by Gurtin (1972). References to further important contributions to the mathematical development and engineering applications of Betti-Maxwell reciprocal

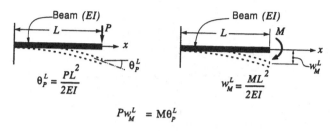

$$Pw_M^L = M\theta_P^L$$

Figure G2 Reciprocal states.

theorems are given by Love (1927), Oden (1967), Timoshenko and Goodier (1970), and Volterra and Gaines (1971). In the ensuing, we shall apply Betti's reciprocal theorem to some problems of special relevance to geomechanics.

(a) The cable-jacking test

The cable-jacking test is a particular configuration of the plate-load test discussed in Section 2.10. It is an in situ test in which a plate resting on the surface of a geological medium is subjected to an axisymmetric load. The load displacement behavior of the plate, usually regarded as relatively rigid, is then utilized to ascertain the elastic properties of the geomaterial. In the cable-jacking test, an anchorage embedded within the geological medium is used as the reaction system. The test procedure is described in detail by Zienkiewicz and Stagg (1967), Jaeger (1972), Roberts (1977), and Bell (1980). The method of application of the load is self-stressing; consequently, no provision is required for reaction weights (Figure G3). In this self-stressing system, there is an interaction between the loaded plate and the embedded anchorage. Ideally, the anchorage must be located at a large distance in comparison to the dimension of the plate, in order to minimize the interaction effects. The extent of this influence needs to be established by examining the problem of the *combined* interaction between the externally loaded plate and the internally loaded half-space (Figure G3). For the purposes of our discussion, we shall examine the case where the rigid plate is subjected to a central load of magnitude $\mathscr{P}$, and the internal anchorage load corresponds to a force of equal magnitude applied at a finite depth ($z = h$) along the vertical axis (Figure G3).

The problem can be examined in two ways. The most direct approach would be to treat the problem as a contact problem in the mathematical theory of elasticity, which examines the indentation of a surface of a half-space internally loaded by a Mindlin force. This approach was applied by Selvadurai (1978) to develop exact closed-form solutions for the combined interaction problem for cases where the internal anchor load is either a concentrated force or is distributed over a finite anchorage length (Selvadurai, 1979).

The alternative approach is to apply Betti's reciprocal theorem. In order to apply the theorem, let us consider two states of loading of the half-space. The first involves Boussinesq's classical problem of the surface loading of a half-space by a rigid circular plate with a smooth contact surface (Figure G4(a)). The second problem involves the internal loading of a half-space region by a Mindlin-type concentrated force, where

(i)   the entire surface of the half-space region is free of shear traction,
(ii) the region corresponding to the test plate area is constrained to deform with a uniform displacement, and
(iii) the region exterior to the test-plate region is free of normal tractions (Figure G4(b)).

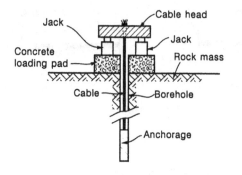

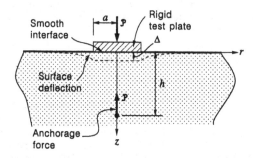

Figure G3 Cable-jacking test, test arrangement, and idealization.

Applying Betti's reciprocal theorem to the two states we have

$$\mathcal{P}(-\Delta^*) = (-\mathcal{P}^*)\Delta \qquad\qquad (G9)$$

where Δ is the displacement in the interior of the half-space at $z = h$, purely due to the test-plate loading (i.e., in the absence of the interior load).

In the cable-jacking test $\mathcal{P} = \mathcal{P}^*$, hence

$$\Delta^* = \Delta. \qquad\qquad (G10)$$

In the case where the test plate is subjected to both an external load $\mathcal{P}$ and the elastic half-space is loaded internally by a load $\mathcal{P}$, the net settlement of the test plate is given by

$$(\Delta)_{\text{Test plate}} = \Delta_o - \Delta^* \qquad\qquad (G11)$$

where Δ_o is the surface settlement of an externally loaded rigid circular plate.

From (G9) and (G10) we have

$$(\Delta)_{\text{Test plate}} = \Delta_o - \Delta. \qquad\qquad (G12)$$

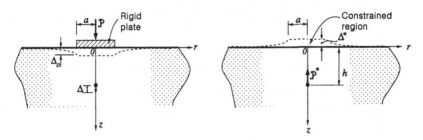

Figure G4 Reciprocal states, (a) indentation of half-space by rigid plate and (b) internal loading of constrained half-space.

From the results given in Chapter 4 we have

$$\Delta_o = \frac{\mathcal{P}(1 - \nu)}{4Ga}.$$ (G13)

The displacement Δ can be obtained by making use of Boussinesq's solution for the concentrated normal force acting on the surface of a half-space and the stress distribution beneath a rigid circular foundation subjected to an axial force $\mathcal{P}$. Considering the contact stress distribution

$$\sigma_{zz}(r, 0) = \frac{2G\Delta_o}{\pi(1 - \nu)} \frac{1}{\sqrt{a^2 - r^2}}; \, 0 < r < a.$$ (G14)

The axial displacement within the half-space region at a depth $(0, z)$, due to the increment of load $\{\sigma_{zz}(r,0)rdrd\theta\}$ is given by

$$u_z(0, z) = \int_0^a \int_0^{2\pi} \frac{\Delta_o r}{2\pi^2(1 - \nu)\sqrt{a^2 - r^2}} \left[\frac{2(1 - \nu)}{(r^2 + z^2)^{1/2}} + \frac{z^2}{(r^2 + z^2)^{3/2}} \right] drd\theta.$$ (G15)

Integrating (G14) we obtain

$$u_z(0, z) = \Delta_o \left[\frac{2}{\pi} \tan^{-1}\left(\frac{a}{z}\right) + \frac{az}{\pi(1 - \nu)(r^2 + z^2)^{3/2}} \right].$$ (G16)

Considering the displacement at $z = h$, $u_z(0, h) = \Delta$, we can evaluate (G11) as follows:

$$(\Delta)_{\text{Test plate}} = \frac{\mathcal{P}(1 - \nu)}{4Ga} \left[1 - \left\{ \frac{2}{\pi} \tan^{-1}\left(\frac{a}{h}\right) + \frac{ah}{\pi(1 - \nu)(a^2 + h^2)} \right\} \right].$$ (G17)

The closed form result (G16) is exactly identical to that obtained by Selvadurai (1978). It is evident that, as $h \to \infty$, $(\Delta)_{\text{Test plate}} \to \Delta_o$ and as $h \to 0$, $(\Delta)_{\text{Test plate}} \to 0$ (i.e., the anchorage load counteracts the externally applied load).

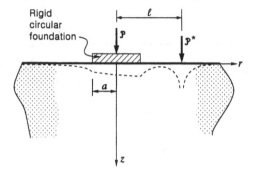

Figure G5 Interaction between a rigid circular foundation and an externally placed load.

(b) Interaction between a rigid circular foundation and an external force

Consider the problem of the interaction between a centrally loaded rigid circular foundation on the surface of a half-space and a concentrated normal force acting on the surface
of the half-space at an exterior point. Due to the presence of the external load $\mathcal{P}^*$, the problem is no longer symmetric about the z-axis. The analysis of the problem can be achieved
by solving a mixed-boundary value problem in which displacements are prescribed within
the foundation region and tractions are prescribed in the remainder of the surface of the
half-space region. The dual integral-equation approach to the solution of this problem is
given by Selvadurai (1980). A result of some engineering interest concerns the additional
settlement of the rigid circular foundation due to the externally applied load $\mathcal{P}^*$. This result can be conveniently evaluated by utilizing Betti's reciprocal theorem.

Consider first the problem of the axisymmetric loading of a rigid circular foundation with a smooth base subjected to a central force $\mathcal{P}$. The surface displacements of
the half-space region are given by (Chapter 4)

$$u_z(r, 0) = \Delta_o = \frac{\mathcal{P}(1 - \nu)}{4Ga}; \; 0 \le r \le a \tag{G18}$$

$$u_z(r, 0) = \Delta_o \left\{ \frac{2}{\pi} \sin^{-1}\left(\frac{a}{r} \right) \right\}; \; a \le r \le \infty. \tag{G19}$$

Hence, referring to Figure G6, the surface settlement at a distance $r = l$ from the
center of the rigid circular foundation is given by

$$\Delta = \Delta_o \left\{ \frac{2}{\pi} \sin^{-1}\left(\frac{a}{l} \right) \right\}. \tag{G20}$$

Now consider the problem where the surface of the half-space region is subjected
to a normal load $\mathcal{P}^*$ applied at a distance $r = l$.

The region of the half-space corresponding to the foundation area, $0 \le r \le a$, is
constrained to deform in a planar fashion, i.e.

$$u_z(r, 0) = \Delta^* + \Omega \, r \cos\theta; \; 0 \le r \le a \tag{G21}$$

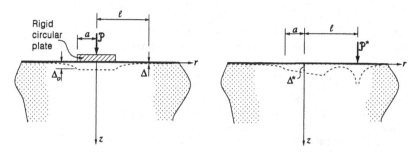

Figure G6 Reciprocal states, (a) axisymmetric indentation of a half-space region and (b) loading of a constrained half-space.

where Ω is the rotation of the constrained region. The entire surface of the half-space region in Figure G6(b) is free of shear tractions, and the region exterior to the constrained region is also free of normal traction. The location $r = l$, at which the concentrated force $\mathscr{P}^*$ acts, exhibits singular behavior in the displacement, strain, and stress fields. Applying Betti's reciprocal theorem to these two states we have

$$\mathscr{P}\,\Delta^* = \mathscr{P}^*\,\Delta. \tag{G22}$$

Hence, the additional displacement Δ^* induced at the center of the rigid circular foundation due to the externally placed load $\mathscr{P}^*$ is given by

$$\Delta^* = \frac{\mathscr{P}^*(1 - \nu)}{2\pi G a} \sin^{-1}\left(\frac{a}{l}\right) \tag{G23}$$

where (G17) and G19) have been used.

It is evident that when $l \to a$, and $\mathscr{P}^* = \mathscr{P}$, $\Delta^* = \Delta_o$, and when $l \to \infty$, $\Delta^* \to 0$.

The preceding application of Betti's reciprocal theorem focused on the evaluation of the displacements at the *center* of the rigid circular foundation due to an external force acting on the surface of the elastic half-space. The procedure can be extended to include the case where the reciprocal theorem can be applied to evaluate the displacements at other locations of the rigid circular foundation due to an externally placed load. Details of these analyses and references to further applications involving distributed external loads are given by Selvadurai (1982).

References

Bell, F. G., *Engineering Geology and Geotechnics*, Newnes-Butterworths, London (1980).

Betti, E., "Teoria dell'Elasticita," *Nuovo Cimento Serie II, VII* and *VIII* (1872).

Gurtin, M. E., "The linear theory of elasticity," In C. Truesdell (Ed.), *Handbuch der Physik*, Band IV a/2, Mechanics of Solids II, Springer-Verlag, Berlin (1972).

Jaeger, C., *Rock Mechanics and Engineering*, Cambridge University Press, Cambridge (1972).

Love, A. E. H., *A Treatise on the Mathematical Theory of Elasticity*, Cambridge University Press, London (1927).

Maxwell, J. C., "On the calculation of the equilibrium and stiffness of frames," *Phil. Magazine*, Vol. 27, p. 294–299 (1864).

Oden, J. T., *Mechanics of Elastic Structures*, McGraw-Hill, New York (1967).

Roberts, A., *Geotechnology: An Introductory Text for Students and Engineers*, Pergamon Press, Oxford (1977).

Selvadurai, A. P. S., "The interaction between a rigid circular punch on an elastic halfspace and a Mindlin force," *Mech. Res. Comm.*, Vol. 5, pp. 57–64 (1978).

Selvadurai, A. P. S., "The displacement of a rigid circular foundation anchored to an isotropic elastic halfspace," *Geotechnique*, Vol. 29, pp. 195–202 (1979).

Selvadurai, A. P. S., "On the displacement induced in a rigid circular punch on an elastic halfspace due to an external force," *Mech. Res. Comm.*, Vol. 7, pp. 351–358 (1980).

Selvadurai, A. P. S., "The additional settlement of a rigid circular foundation on an isotropic elastic halfspace due to multiple distributed external loads," *Geotechnique*, Vol. 32, pp. 1–7 (1982).

Timoshenko, S. P., and Goodier, J. N., *Theory of Elasticity*, McGraw-Hill, New York (1970).

Volterra, E., and Gaines, J. H., *Advanced Strength of Materials*, Prentice-Hall, New Jersey (1971).

Zienkiewicz, O. C., and Stagg, K. G., "Cable method of in situ testing," *Int. J. Rock Mech. Min. Sci.*, Vol. 4, pp. 273–300 (1967).

Index

Index